A PRIMER OF QUATERNIONS

ARTHUR S. HATHAWAY
PROFESSOR OF MATHEMATICS IN THE ROSE POLYTECHNIC
INSTITUTE, TERRE HAUTE, IND.

HAWK PRESS

Published by

Hawk Press
4836/24, Ansari Road, Daryaganj
New Delhi – 110 002
Phones : +91-11-23278618, +91-11-43667199
E-mail : thehawkpress@gmail.com
www.thehawkpress.com

ISBN : 978-93-93971-55-5

Preface

The Theory of Quaternions is due to Sir William Rowan Hamilton, Royal Astronomer of Ireland, who presented his first paper on the subject to the Royal Irish Academy in 1843. His Lectures on Quaternions were published in 1853, and his Elements, in 1866, shortly after his death. The Elements of Quaternions by Tait is the accepted text-book for advanced students.

The following development of the theory is prepared for average students with a thorough knowledge of the elements of algebra and geometry, and is believed to be a simple and elementary treatment founded directly upon the fundamental ideas of the subject. This theory is applied in the more advanced examples to develop the principal formulas of trigonometry and solid analytical geometry, and the general properties and classification of surfaces of second order.

In the endeavour to bring out the *number* idea of Quaternions, and at the same time retain the established nomenclature of the analysis, I have found it necessary to abandon the term "*vector*" for a directed length. I adopt instead Clifford's suggestive name of "*step*," leaving to "*vector*" the sole meaning of "*right quaternion*." This brings out clearly the relations of this number and line, and emphasizes the fact that Quaternions is a natural extension of our fundamental ideas of number, that is subject to ordinary principles of geometric representation, rather than an artificial species of geometrical algebra.

The physical conceptions and the breadth of idea that the subject of Quaternions will develop are, of themselves, sufficient reward for its study. At the same time, the power, directness, and simplicity of its analysis cannot fail to prove useful in all physical and geometrical investigations, to those who have thoroughly grasped its principles.

On account of the universal use of analytical geometry, many examples have been given to show that Quaternions in its semi-cartesian form is a direct development of that subject. In fact, the present work is the outcome of lectures that I have given to my classes for a number of years past as the equivalent of the usual instruction in the analytical geometry of space. The main features of this primer were therefore developed in the laboratory of the class-room, and I desire to express my thanks to the members of my classes, wherever they may be, for the interest that they have shown, and the readiness with which they have expressed their difficulties, as it has been a constant source of encouragement and assistance in my work.

I am also otherwise indebted to two of my students,—to Mr. H. B. Stilz for the accurate construction of the diagrams, and to Mr. G. Willius for the plan (upon the cover) of the plagiograph or mechanical quaternion multiplier which was made by him while taking this subject. The theory of this instrument is contained in the step proportions that are given with the diagram.[1]

ARTHUR S. HATHAWAY.

[1]See Example 19, Chapter I.

Contents

Chapter 1

Steps

1. Definition. *A step is a given length measured in a given direction.*

   E.g., 3 feet east, 3 feet north, 3 feet up, 3 feet north-east, 3 feet north-east-up, are steps.

2. Definition. *Two steps are equal when, and only when, they have the same lengths and the same directions.*

   E.g., 3 feet east, and *3 feet north,* are not equal steps, because they differ in direction, although their lengths are the same; and *3 feet east, 5 feet east,* are not equal steps, because their lengths differ, although their directions are the same; but all steps of *3 feet east* are equal steps, whatever the points of departure.

3. We shall use bold-faced **AB** to denote the step whose length is AB, and whose direction is from A towards B.

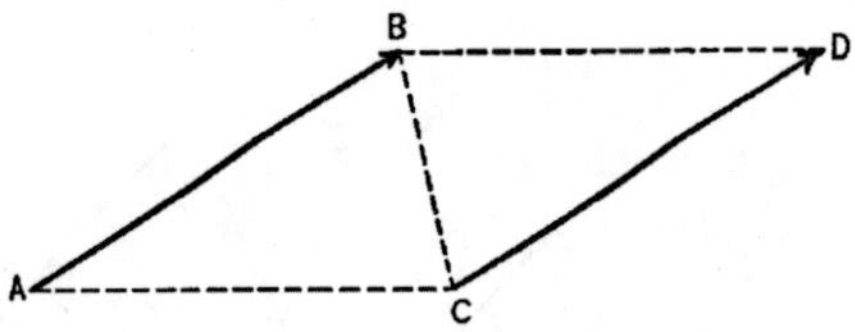

   Two steps **AB**, **CD**, are obviously equal when, and only when, $ABDC$ is a parallelogram.

4. Definition. *If several steps be taken in succession, so that each step begins where the preceding step ends, the step from the beginning of the first to the end of the last step is the sum of those steps.*

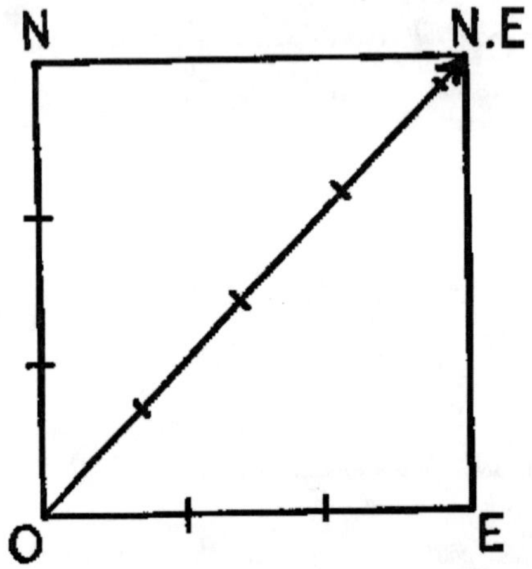

E.g., 3 feet east + 3 feet north $= 3\sqrt{2}$ *feet north-east = 3 feet north + 3 feet east.* Also $\mathbf{AB} + \mathbf{BC} = \mathbf{AC}$, whatever points A, B, C, may be. Observe that this equality between *steps* is not a length equality, and therefore does not contradict the inequality $AB + BC > AC$, just as 5 *dollars credit* + 2 *dollars debit* = 3 *dollars credit* does not contradict the inequality *5 dollars + 2 dollars > 3 dollars.*

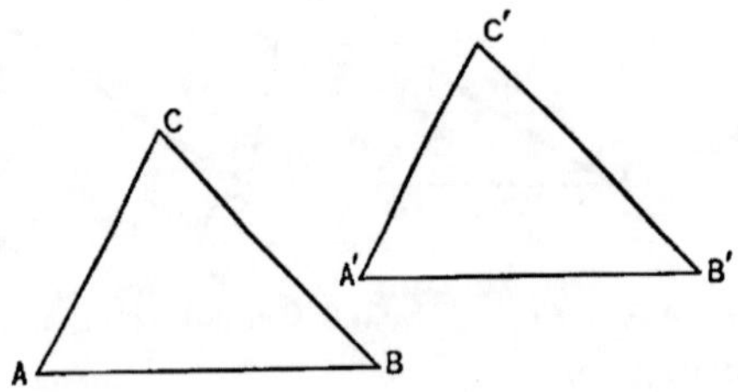

5. *If equal steps be added to equal steps, the sums are equal steps.*

Thus if $\mathbf{AB} = \mathbf{A'B'}$, and $\mathbf{BC} = \mathbf{B'C'}$, then $\mathbf{AC} = \mathbf{A'C'}$, since the triangles ABC, $A'B'C'$ must be equal triangles with the corresponding sides in the same direction.

6. *A sum of steps is commutative* (*i.e.*, the components of the sum may be added in any order without changing the value of the sum).

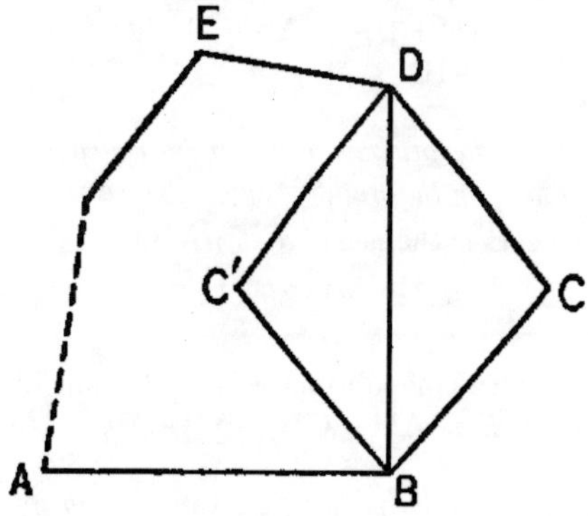

For, in the sum $\mathbf{AB} + \mathbf{BC} + \mathbf{CD} + \mathbf{DE} + \cdots$, let $\mathbf{BC'} = \mathbf{CD}$; then since $BCDC'$ is a parallelogram, therefore $\mathbf{C'D} = \mathbf{BC}$, and the sum with $\mathbf{BC}$, $\mathbf{CD}$, interchanged is $\mathbf{AB} + \mathbf{BC'} + \mathbf{C'D} + \mathbf{DE} + \cdots$, which has the same value as before. By such interchanges, the sum can be brought to any order of adding.

7. *A sum of steps is associative* (*i.e.*, any number of consecutive terms of the sum may be replaced by their sum without changing the value of the whole sum).

For, in the sum $\mathbf{AB} + \mathbf{BC} + \mathbf{CD} + \mathbf{DE} + \cdots$, let $\mathbf{BC}$, $\mathbf{CD}$, be replaced by their sum $\mathbf{BD}$; then the new sum is $\mathbf{AB} + \mathbf{BD} + \mathbf{DE} + \cdots$, whose value is the same as before; and similarly for other consecutive terms.

8. *The product of a step by a positive number is that step lengthened by the multiplier without change of direction.*

E.g., $\mathbf{2AB} = \mathbf{AB} + \mathbf{AB}$, which is $\mathbf{AB}$ doubled in length without change of direction; similarly $\frac{1}{2}\mathbf{AB}$ =(step that doubled gives $\mathbf{AB}$) = ($\mathbf{AB}$ halved in length without change of direction). In general, $m\mathbf{AB} = m$ lengths AB measured in the direction $\mathbf{AB}$; $\frac{1}{n}\mathbf{AB} = \frac{1}{n}$th of length AB measured in the direction $\mathbf{AB}$; etc.

9. *The negative of a step is that step reversed in direction without change of length.*

   For the negative of a quantity is that quantity which added to it gives zero; and since $\mathbf{AB} + \mathbf{BA} = \mathbf{AA} = 0$, therefore $\mathbf{BA}$ is the negative of $\mathbf{AB}$, or $\mathbf{BA} = -\mathbf{AB}$.

   - COR. 1. *The product of a step by a negative number is that step lengthened by the number and reversed in direction.*

     For $-n\mathbf{AB}$ is the negative of $n\mathbf{AB}$.
   - COR. 2. *A step is subtracted by reversing its direction and adding it.*

     For the result of subtracting is the result of adding the negative quantity. *E.g.*, $\mathbf{AB} - \mathbf{CB} = \mathbf{AB} + \mathbf{BC} = \mathbf{AC}$.

10. *A sum of steps is multiplied by a given number by multiplying the components of the sum by the number and adding the products.*

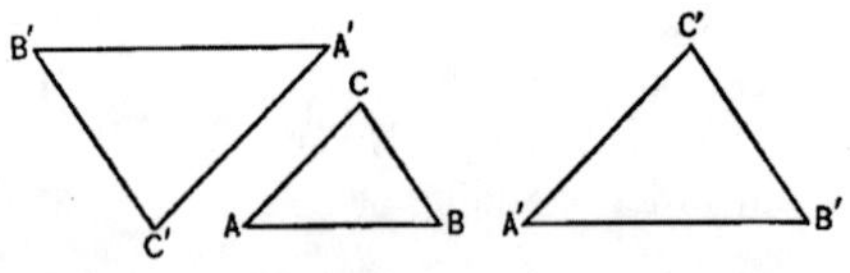

Let $n \cdot \mathbf{AB} = \mathbf{A'B'}, n \cdot \mathbf{BC} = \mathbf{BC'}$; then $ABC, A'B'C'$ are similar triangles, since the sides about B, B' are proportional, and in the same or opposite directions, according as n is positive or negative; therefore AC, $A'C'$ are in the same or opposite directions and in the same ratio; *i.e.*, $n\mathbf{AC} = \mathbf{A'C'}$, which is the same as $n(\mathbf{AB} + \mathbf{BC}) = n\mathbf{AB} + n\mathbf{BC}$.

This result may also be stated in the form: *a multiplier is distributive over a sum.*

11. *Any step may be resolved into a multiple of a given step parallel to it; and into a sum of multiples of two given steps in the same plane with it that are not parallel; and into a sum of multiples of three given steps that are not parallel to one plane.*

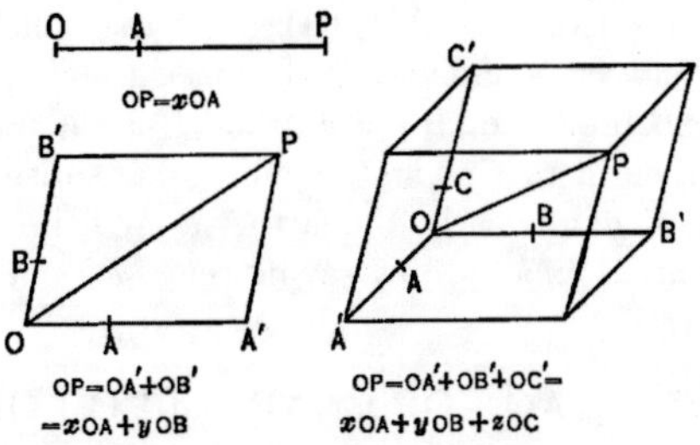

12. It is obvious that if the sum of two finite steps is zero, then the two steps must be parallel; in fact, if one step is **AB**, then the other must be equal to **BA**. Also, if the sum of three finite steps is zero, then the three steps must be parallel to one plane; in fact, if the first is **AB**, and the second is **BC**, then the third must be equal to **CA**. Hence, *if a sum of steps on two lines that are not parallel (or on three lines that are not parallel to one plane) is zero, then the sum of the steps on each line is zero,* since, as just shown, the sum of the steps on each line cannot be finite and satisfy the condition that their sum is zero. We thus see that an equation between steps of one plane can be separated into two equations by resolving each step parallel to two intersecting lines of that plane, and that an equation between steps in space can be separated into three equations by resolving each step parallel to three lines of space that are not parallel to one plane. We proceed to give some applications of this and other principles of step analysis in locating a point or a locus of points with respect to given data (Arts. 13-20).

Centre of Gravity

13. *The point P that satisfies the condition $l\mathbf{AP} + m\mathbf{BP} = 0$ lies upon the line AB and divides AB in the inverse ratio of $l : m$ (i.e., P is the*

centre of gravity of a mass l at A and a mass m at B).

The equation gives l**AP** $= m$**PB**; hence:

AP, **PB** are parallel; P lies on the line AB; and **AP** : **PB** $= m : l =$ *inverse of* $l : m$.

If $l : m$ is positive, then **AP**, **PB** are in the same direction, so that P must lie between A and B; and if $l : m$ is negative, then P must lie on the line AB produced. If $l = m$, then P is the middle point of AB; if $l = -m$, then there is no finite point P that satisfies the condition, but P satisfies it more nearly, the farther away it lies upon AB produced, and this fact is expressed by saying that *"P is the point at infinity on the line AB."*

14. By substituting **AO** + **OP** for **AP** and **BO** + **OP** for **BP** in l**AP** + m**BP** $= 0$, and transposing known steps to the second member, we find the point P with respect to any given origin O, viz.,

(a) $(l+m)$**OP** $= l$**OA** $+ m$**OB**, where P divides AB inversely as $l : m$.

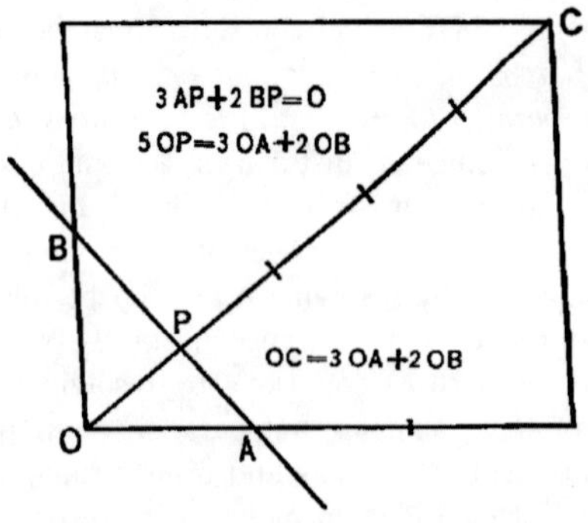

- Cor. *If* **OC** $= l$**OA** $+ m$**OB**, *then OC, produced if necessary, cuts AB in the inverse ratio of* $l : m$, *and* **OC** *is* $(l + m)$ *times the step from O to the point of division.*

  For, if P divide AB inversely as $l : m$, then by *(a)* and the given equation, we have

$$\mathbf{OC} = (l + m)\mathbf{OP}.$$

15. *The point P that satisfies the condition $l\mathbf{AP} + m\mathbf{BP} + n\mathbf{CP} = 0$ lies in the plane of the triangle ABC; AP (produced) cuts BC at a point D that divides BC inversely as $m : n$, and P divides AD inversely as $l : m+n$ (i.e., P is the center of gravity of a mass l at A, a mass m at B, and a mass n at C). Also the triangles PBC, PCA, PAB, ABC, are proportional to l, m, n, $l+m+n$.*

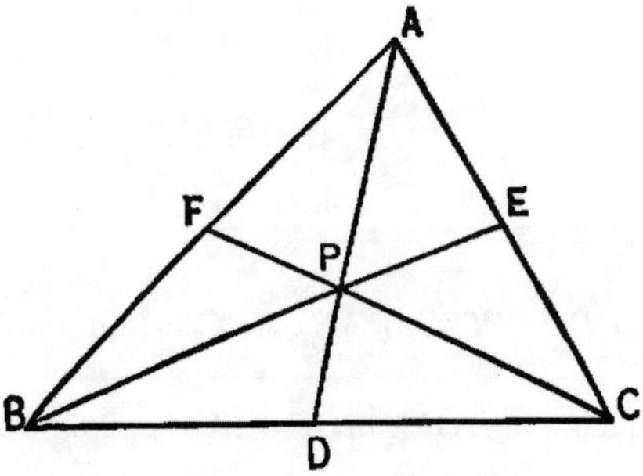

The three steps $l\mathbf{AP}$, $m\mathbf{BP}$, $n\mathbf{CP}$ must be parallel to one plane, since their sum is zero, and hence P must lie in the plane of ABC. Since $\mathbf{BP} = \mathbf{BD} + \mathbf{DP}$, $\mathbf{CP} = \mathbf{CD} + \mathbf{DP}$, the equation becomes, by making these substitutions, $l\mathbf{AP} + (m+n)\mathbf{DP} + m\mathbf{BD} + n\mathbf{CD} = 0$. This is an equation between steps on the two intersecting lines, AD, BC, and hence the resultant step along each line is zero; i.e., $m\mathbf{BD} + n\mathbf{CD} = 0$ (or D divides BC inversely as $m : n$), and

(a) $$l\mathbf{AP} + (m+n)\mathbf{DP} = 0$$

(or P divides AD inversely as $l : m+n$). Also, we have, by adding $l\mathbf{PD} + l\mathbf{DP} = 0$ to (a),

$$l\mathbf{AD} + (l+m+n)\mathbf{DP} = 0.$$

Hence

$$l : l+m+n = \mathbf{PD} : \mathbf{AD} = PBC : ABC,$$

since the triangles PBC, ABC have a common base BC, (We must take the ratio of these triangles as positive or negative according as the vertices P, A lie on the same or opposite sides of the base BC, since the ratio $\mathbf{PD} : \mathbf{AD}$ is positive or negative under those circumstances.) Similarly,

$$PCA : ABC = m : l + m + n,$$

and

$$PAB : ABC = n : l + m + n.$$

Hence, we have,

$$PBC : PCA : PAB : ABC = l : m : n : l + m + n.$$

16. By introducing in $l\mathbf{AP} + m\mathbf{BP} + n\mathbf{CP} = 0$ an origin O, as in Art. 14, we find

(*a*) $(l + m + n)\mathbf{OP} = l\mathbf{OA} + m\mathbf{OB} + n\mathbf{OC}$, *where P divides ABC in the ratio $l : m : n$.*

NOTE. As an exercise, extend this formula for the center of gravity P, of masses l, m, n, at A, B, C, to four or more masses.

Curve Tracing. Tangents.

17. *To draw the locus of a point P that varies according to the law* $\mathbf{OP} = t\mathbf{OA} + \frac{1}{2}t^2\mathbf{OB}$, *where t is a variable number.* (*E.g.*, t = number of seconds from a given epoch.)

Take $t = -2$, and P is at D', where

$$\mathbf{OD'} = -2\mathbf{OA} + 2\mathbf{OB}.$$

Take $t = -1$, and P is at C', where

$$\mathbf{OC'} = -\mathbf{OA} + \frac{1}{2}\mathbf{OB}$$

Take $t = 0$, and P is at O. Take $t = 1$, and P is at C, where $\mathbf{OC} = \mathbf{OA} + \frac{1}{2}\mathbf{OB}$. Take $t = 2$, and P is at D, where $\mathbf{OD} = 2\mathbf{OA} + 2\mathbf{OB}$. It is thus seen that when t varies from -2 to 2, then P traces a curve $D'C'OCD$. To draw the curve as accurately as possible, we find the tangents at the points already found. The method that we employ is perfectly general and applicable to any locus.

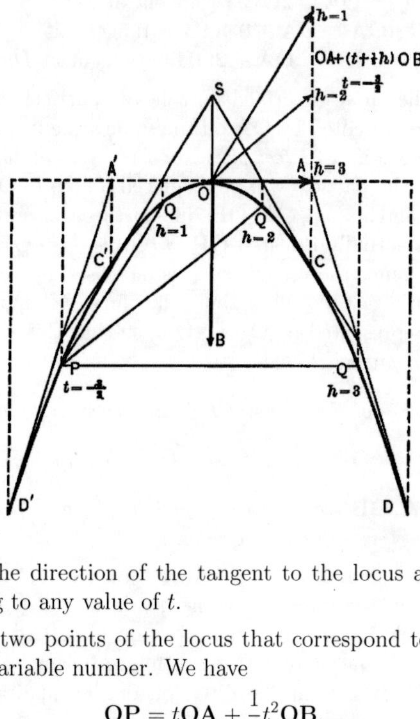

(a) To find the direction of the tangent to the locus at the point P corresponding to any value of t.

Let P, Q be two points of the locus that correspond to the values t, $t + h$ of the variable number. We have

$$\mathbf{OP} = t\mathbf{OA} + \frac{1}{2}t^2\mathbf{OB},$$

$$\mathbf{OQ} = (t + h)\mathbf{OA} + \frac{1}{2}(t + h)^2\mathbf{OB},$$

and therefore

$$\mathbf{PQ} = \mathbf{OQ} - \mathbf{OP} = h\left[\mathbf{OA} + (t + \frac{1}{2}h)\mathbf{OB}\right].$$

Hence (dropping the factor h) we see that $\mathbf{OA} + (t + \frac{1}{2}h)\mathbf{OB}$ is always *parallel* to the chord PQ. Make h approach 0, and then Q approaches P, and the (indefinitely extended) chord PQ approaches coincidence with the tangent at P. Hence making $h = 0$, in the step that is parallel to the chord, we find that $\mathbf{OA} + t\mathbf{OB}$ is parallel to the tangent at P.

Apply this result to the special positions of P already found, and we have: $\mathbf{D'A'} = \mathbf{OA} - 2\mathbf{OB}$ = tangent at D'; $\mathbf{C'S} = \mathbf{OA} - \mathbf{OB}$ = tangent at C'; $\mathbf{OA} = \mathbf{OA} + \mathbf{0} \cdot \mathbf{OB}$ = tangent at O; $\mathbf{SO} = \mathbf{OA} + \mathbf{OB}$ = tangent at C; $\mathbf{AD} = \mathbf{OA} + 2\mathbf{OB}$ = tangent at D.

This is the curve described by a heavy particle thrown from O with velocity represented by $\mathbf{OA}$ on the same scale in which $\mathbf{OB}$ represents an acceleration of 32 *feet per second per second downwards.* For, after t seconds the particle will be displaced a step $t \cdot \mathbf{OA}$ due to its initial velocity, and a step $\frac{1}{2}t^2 \cdot \mathbf{OB}$ due to the acceleration downwards, so that P is actually the step $\mathbf{OP} = t\mathbf{OA} + \frac{1}{2}t^2 \cdot \mathbf{OB}$ from O at time t. Similarly, since the velocity of P is increased by a velocity represented by $\mathbf{OB}$ in every second of time, therefore P is moving at time t with velocity represented by $\mathbf{OA} + t\mathbf{OB}$, so that this step must be parallel to the tangent at P.

18. *To draw the locus of a point P that varies according to the law*

$$\mathbf{OP} = \cos(nt + e) \cdot \mathbf{OA} + \sin(nt + e) \cdot \mathbf{OB},$$

where $\mathbf{OA}, \mathbf{OB}$ *are steps of equal length and perpendicular to each other, and t is any variable number.*

With centre O and radius OA draw the circle $ABA'B'$. Take arc $AE = e$ radians in the direction of the quadrant AB (*i.e.* an arc of e radii of the circle in length in the direction of AB or AB' according as e is positive or negative). Corresponding to any value of t, lay off arc $EP = nt$ radians in the direction of the quadrant AB. Then arc $AP = nt + e$ radians. Draw LP perpendicular to OA at L. Then according to the definitions of the trigonometric functions of an angle we have,

$$\cos(nt + e) = \overline{OL}/OP, \quad \sin(nt + e) = \overline{LP}/OP.$$ [1]

[1]Observe the distinctions: $\mathbf{OL}$, a step; $\overline{OL}$, a positive or negative length of a directed axis; OL, a length.

Hence we have for all values of t,

$$\mathbf{OL} = \cos(nt + e)\mathbf{OA}, \quad \mathbf{LP} = \sin(nt + e)\mathbf{OB},$$

and adding these equations, we find that

$$\mathbf{OP} = \cos(nt + e)\mathbf{OA} + \sin(nt + e)\mathbf{OB}.$$

Hence, *the locus of the required point P is the circle on* $\mathbf{OA}, \mathbf{OB}$ *as radii.*

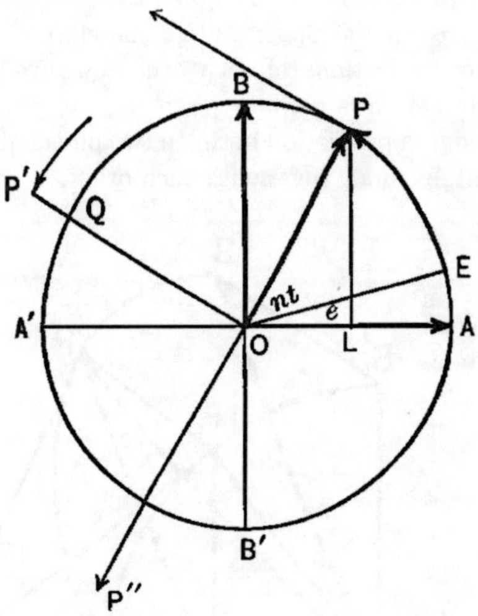

Let t be the number of seconds that have elapsed since epoch. Then, at epoch, $t = 0$, and P is at E; and since in t seconds P has moved through an arc EP of nt radians, therefore P moves uniformly round the circle at the rate of n radians per second. Its velocity at time t is therefore represented by n times that radius of the circle which is perpendicular to OP in the direction of its motion, or by $\mathbf{OP}' = n\mathbf{OQ}$, where arc $PQ = \frac{\pi}{2}$ radians. Hence, since arc $AQ = (nt+e+\frac{\pi}{2})$ radians, therefore $\mathbf{OP}' = n\left[\cos\left(nt + e + \frac{\pi}{2}\right) \cdot \mathbf{OA} + \sin\left(nt + e + \frac{\pi}{2}\right) \cdot \mathbf{OB}\right]$. The point $P\prime$ also moves uniformly in a circle, and this circle is the hodograph of the motion. The velocity in the hodograph (or the acceleration of P) is similarly $\mathbf{OP}'' = n^2\mathbf{PO}$.

Parallel Projection

19. *If* $\mathbf{OP} = x\mathbf{OA} + y\mathbf{OB}$, $\mathbf{OP'} = x\mathbf{OA} + y\mathbf{OB'}$, *where* x, y *vary with the arbitrary number* t *according to any given law so that* P, P' *describe definite loci (and have definite motions when* t *denotes time), then the two loci (and motions) are parallel projections of each other by rays that are parallel to* BB',

For, by subtracting the two equations we find $\mathbf{PP'} = y\mathbf{BB'}$, so that PP' is always parallel to BB'; and as P moves in the plane AOB and P' moves in the plane AOB', therefore their loci (and motions) are parallel projections of each other by rays parallel to BB'. The parallel projection is definite when the two planes coincide, and may be regarded as a projection between two planes AOB, AOB', that make an indefinitely small angle with each other.

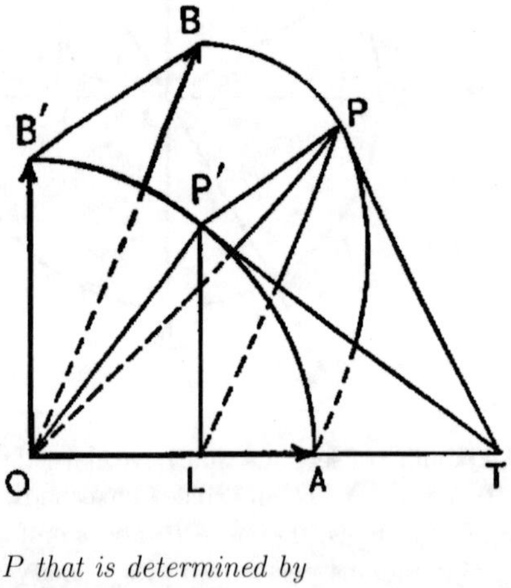

20. *The motion of* P *that is determined by*

$$\mathbf{OP} = \cos(nt + e)\mathbf{OA} + \sin(nt + e)\mathbf{OB}$$

is the parallel projection of uniform circular motion.

For, draw a step $\mathbf{OB'}$ perpendicular to $\mathbf{OA}$ and equal to it in length. Then, by Art. 18, the motion of P' determined by

$$\mathbf{OP'} = \cos(nt + e)\mathbf{OA} + \sin(nt + e)\mathbf{OB'}$$

is a uniform motion in a circle on **OA**, **OB′** as radii; and by Art. 19 this is in parallel perspective with the motion of P.

Step Proportion

21. Definition. *Four steps* **AC**, **AB**, **A′C′**, **A′B′** *are in proportion when the first is to the second in respect to both relative length and relative direction as the third is to the fourth in the same respects.*

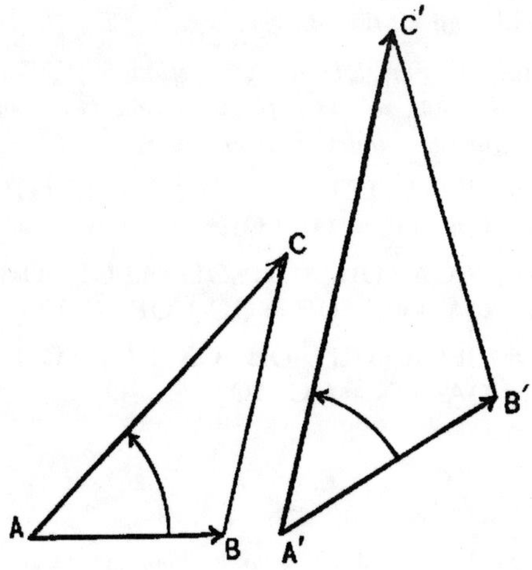

This requires, first, that the lengths of the steps are in proportion or

$$AC : AB = A'C' : A'B';$$

and secondly, that **AC** deviates from **AB** by the same plane angle in direction and magnitude that **A′C′** deviates from **A′B′**.

Hence, first, the triangles ABC, $A'B'C'$ are similar, since the angles A, A' are equal and the sides about those angles are proportional; and secondly, one triangle may be turned in its plane into a position in which its sides lie in the same directions as the corresponding sides of the other triangle. Two such triangles will be called *similar and congruent triangles*, and corresponding angles will be called *congruent angles*.

22. We give the final propositions of Euclid, Book V., as exercises in step proportion.

    - (xi.) *If four steps are proportionals, they are also proportionals when taken alternately.*
    - (xii.) *If any number of steps are proportionals, then as one of the antecedents is to its consequent, so is the sum of the antecedents to the sum of the consequents.*
    - (xiii.) *If four steps are proportionals, the sum (or difference) of the first and second is to the second as the sum (or difference) of the third and fourth is to the fourth.*
    - (xiv.) If $\mathbf{OA} : \mathbf{OB} = \mathbf{OP} : \mathbf{OQ}$ and $\mathbf{OB} : \mathbf{OC} = \mathbf{OQ} : \mathbf{OR}$, then $\mathbf{OA} : \mathbf{OC} = \mathbf{OP} : \mathbf{OR}$.
    - (xv.) If $\mathbf{OA} : \mathbf{OB} = \mathbf{OC} : \mathbf{OD}$ and $\mathbf{OE} : \mathbf{OB} = \mathbf{OF} : \mathbf{OD}$, then $\mathbf{OA} + \mathbf{OE} : \mathbf{OB} = \mathbf{OC} + \mathbf{OF} : \mathbf{OD}$.
    - (xvi.) If $\mathbf{OA} : \mathbf{OB} = \mathbf{OB} : \mathbf{OX} = \mathbf{OC} : \mathbf{OD} = \mathbf{OD} : \mathbf{OY}$, then $\mathbf{OA} : \mathbf{OX} = \mathbf{OC} : \mathbf{IO}$.

Examples

We shall use $\mathbf{i}$, $\mathbf{j}$, $\mathbf{k}$, as symbols for *unit length east*, *unit length north*, and *unit length up*, respectively.

1. Mark the points whose steps from a given point are $\mathbf{i} + 2\mathbf{j}$, $-3\mathbf{i} - \mathbf{j}$. Show that the step from the first point to the second is $-4\mathbf{i} - 3\mathbf{j}$, and that the length is 5.

2. Show that the four points whose steps from a given point are $2\mathbf{i}+\mathbf{j}$, $5\mathbf{i}+4\mathbf{j}$, $4\mathbf{i}+7\mathbf{j}$, $\mathbf{i}+4\mathbf{j}$ are the angular points of a parallelogram. Also determine their centre of gravity, with weights $1, 1, 1, 1$; also with weights $1, 2, 3, 4$; also with weights $1, -2, 3, -4$.

3. If $\mathbf{OA} = \mathbf{i}+2\mathbf{j}$, $\mathbf{OB} = 4\mathbf{i}+3\mathbf{j}$, $\mathbf{OC} = 2\mathbf{i}+3\mathbf{j}$, $\mathbf{OD} = 4\mathbf{i}+\mathbf{j}$, find CD as sums of multiples of $\mathbf{CA}$, $\mathbf{CB}$, and show that $\mathbf{CD}$ bisects $\mathbf{AB}$.

4. If $\mathbf{OP} = x\mathbf{i} + y\mathbf{j}$, $\mathbf{OP'} = x'\mathbf{i} + y'\mathbf{j}$, then $\mathbf{PP'} = (x' - x)\mathbf{i} + (y' - y)\mathbf{j}$ and $\overline{PP'}^2 = (x' - x)^2 + (y' - y)^2$.

5. Show that $\mathbf{AB}$ is bisected by $\mathbf{OC} = \mathbf{OA} + \mathbf{OB}$, and trisected by $\mathbf{OD} = 2\mathbf{OA} + \mathbf{OB}$, $\mathbf{OE} = \mathbf{OA} + 2\mathbf{OB}$, and divided inversely as $2 : 3$ by $\mathbf{OF} = 2\mathbf{OA} + 3\mathbf{OB}$.

6. Show that $\mathbf{AA'} + \mathbf{BB'} = 2\mathbf{MM'}$, where MM' are the middle points of AB, $A'B'$, respectively.

7. Show that $2\mathbf{AA'} + 3\mathbf{BB'} = (2 + 3)\mathbf{CC'}$, where C, C' are the points that divide AB, $A'B'$, inversely as $2 : 3$. Similarly, when 2, 3 are replaced by l, m.

8. Show that the point that divides a triangle into three equal triangles is the intersection of the medial lines of the triangle.

9. Show that the points which divide a triangle into triangles of equal magnitude, one of which is negative (the given triangle being positive), are the vertices of the circumscribing triangle with sides parallel to the given triangle.

10. If a, b, c are the lengths of the sides BC, CA, AB of a triangle, show that $\frac{1}{b}\mathbf{AC} \pm \frac{1}{c}\mathbf{AB}$ (drawn from A) are interior and exterior bisectors of the angle A; and that when produced they cut the opposite side BC in the ratio of the adjacent sides.

11. The $\left\{\begin{matrix}\text{lines}\\ \text{points}\end{matrix}\right.$ that join the $\left\{\begin{matrix}\text{vertices}\\ \text{sides}\end{matrix}\right.$ of a triangle ABC to any $\left\{\begin{matrix}\text{point } P\\ \text{line } p\end{matrix}\right.$ in its plane divide the sides BC, CA, AB in ratios whose product is $\left\{\begin{matrix}+1\\ -1\end{matrix}\right.$; and conversely $\left\{\begin{matrix}\text{lines from}\\ \text{points on}\end{matrix}\right.$ the $\left\{\begin{matrix}\text{vertices}\\ \text{sides}\end{matrix}\right.$ that so divide the sides $\left\{\begin{matrix}\text{meet in a point.}\\ \text{lie in a line.}\end{matrix}\right.$

12. Prove by Exs. 10, 11, that the three interior bisectors of the angles of a triangle (also an interior and two exterior bisectors) meet in a point; and that the three exterior bisectors (also an exterior and two interior bisectors) meet the sides in colinear points.

13. Determine the locus (and motion) of P, given by $\mathbf{OP} = \mathbf{OA} + t\mathbf{OB}$; also of $\mathbf{OP} = (1 + 2t)\mathbf{i} + (3t - 2)\mathbf{j}$.

14. Compare the loci of P determined by the following *pairs* of **step** and **length** equations:

$$\mathbf{AP} = 2 \text{ east}, \quad AP = 2; \qquad \mathbf{AP} = 2\mathbf{BP}, \quad AP = 2BP;$$
$$\mathbf{AP} + \mathbf{BP} = \mathbf{CD}, \quad AP + BP = CD$$

15. Draw, by points and tangents, the locus of P determined by each of the following values of **OP**, in which x is any number:

$$x\mathbf{i} + \frac{1}{2}x^2\mathbf{j}; \quad x\mathbf{i} + \frac{2}{x}\mathbf{j}; \quad x\mathbf{i} + \frac{1}{3}x^3\mathbf{j}; \quad x\mathbf{i} + (\frac{1}{3}x^3 - x^2 + 2)\mathbf{j};$$
$$x\mathbf{i} + \frac{8}{x^2+4}\mathbf{j}; \quad x\mathbf{i} + \sqrt{4 - x^2}\mathbf{j}; \quad x\mathbf{i} + \frac{1}{2}\sqrt{4 - x^2}\mathbf{j}.$$

16. Take three equal lengths making angles 120° with each other as projections of **i**, **j**, **k**, and construct by points the projection of the locus of P, where $\mathbf{OP} = 2(\cos x \cdot \mathbf{i} + \sin x \cdot \mathbf{j}) + x \cdot \mathbf{k}$, x varying from 0 to 2π. Show that this curve is one turn of a helix round a vertical cylinder of altitude 2π, the base being a horizontal circle of radius 2 round O as centre.

17. A circle rolls inside a fixed circle of twice its diameter; show that any point of the plane of the rolling circle traces a parallel projection of a circle.

18. A plane carries two pins that slide in two fixed rectangular grooves; show that any point of the sliding plane traces a parallel projection of a circle.

19. $OACB$ is a parallelogram whose sides are rigid and jointed so as to turn round the vertices of the parallelogram; APC, BCQ are rigid similar and congruent triangles. Show that $\mathbf{AC} : \mathbf{AP} = \mathbf{BQ} : \mathbf{BC} = \mathbf{OQ} : \mathbf{OP}$, and that therefore P, Q trace similar congruent figures when O remains stationary (21, 22, xii.). [See cover of book.]

20. If the plane pencil OA, OB, OC, OD is cut by any straight line in the points P, Q, B, S, show that the *cross-ratio* $(\overline{PR} : \overline{RQ}) : (\overline{PS} : \overline{SQ})$ is constant for all positions of the line.

$$[\mathbf{OC} = l\mathbf{OA} + m\mathbf{OB} = lx\mathbf{OP} + my\mathbf{OQ} \text{ gives } \overline{PR} : \overline{RQ} = my : lx].$$

21. Two roads run north, and east, intersecting at O. A is 60 *feet south* of O, walking 3 *feet per second north*, B is 60 *feet west* of O, walking 4 *feet per second east*. When are A, B nearest together, and what is B's apparent motion as seen by A?

22. What is B's motion relative to A in Ex. 21 if B is accelerating his walk at the rate of 3 *inches per second per second*?

23. In Ex. 21, let the east road be 20 feet above the level of the north road; and similarly in Ex. 22.

24. A massless ring P is attached to several elastic strings that pass respectively through smooth rings at A, B, C, $\cdots$ and are attached to fixed points A', B', C', $\cdots$ such that $A'A$, $B'B$, $C'C$, $\cdots$ are the natural lengths of the strings. The first string has a tension l per unit of length that it is stretched (Hooke's law), the second a tension m, the third a tension n, etc. Find the resultant force on P and its position of equilibrium.

25. The same as Ex. 24, except that the ring has a mass w.

Chapter 2

Rotations. Turns. Arc Steps

23. Definitions of Rotation

A step is **rotated** when it is revolved about an axis through its initial point as a rigid length rigidly attached to the axis. The step describes a conical angle about the axis except when it is perpendicular to the axis.

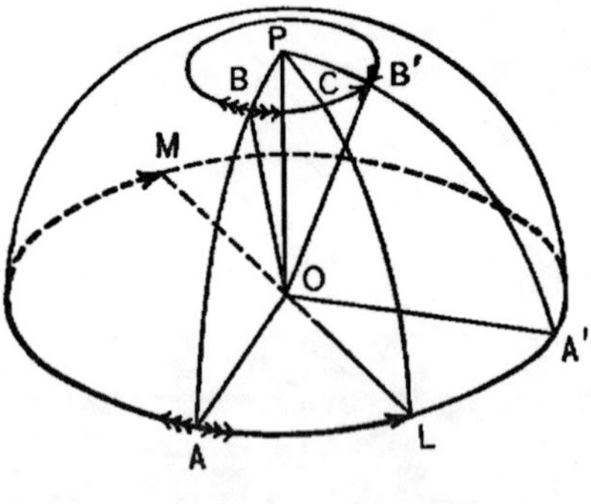

If a rotation through a diedral angle of given magnitude and direction in space be applied to the radii of a sphere of unit radius and centre O, the sphere is rotated as a rigid body about a certain diameter PP' as axis, and a plane through O perpendicular to the axis intersects the sphere in the **equator** of the rotation.

Either of the two directed arcs of the equator from the initial position A to the final position A' of a point of the rotated sphere that lies on the equator is the **arc** of the rotation. If these two arcs he bisected at L, M respectively, then the two arcs are $2\overset{\frown}{AL}$, $2\overset{\frown}{AM}$ respectively, and $\overset{\frown}{AL}$, $\overset{\frown}{AM}$ are supplementary arcs in opposite directions, each less than a semicircle. When these half-arcs are 0° and 180° respectively, they represent a rotation of the sphere into its original position, whose axis and equator are indeterminate, so that such arcs may be measured on any great circle of the sphere without altering the corresponding rotation.

24. *A rotation is determined by the position into which it rotates two given non-parallel steps.*

For let the radii **OB**, **OC** rotate into the radii **OB′**, **OC′**. Any axis round which **OB** rotates into **OB′** must be equally inclined to these radii; *i.e.*, it is a diameter of the great circle PKL that bisects the great arc $\overset{\frown}{BB'}$ at right angles.

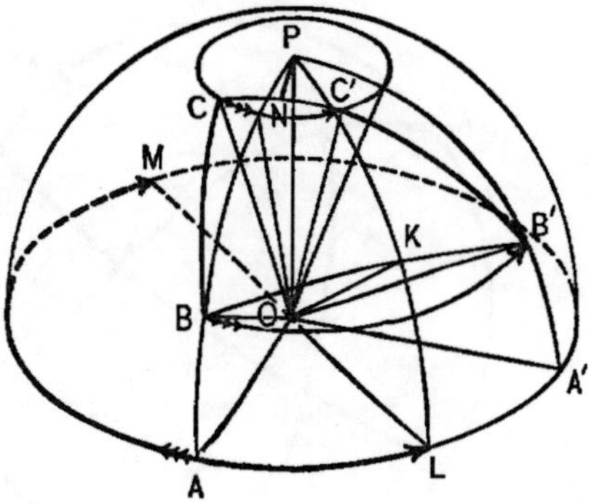

E.g., OK, OL, OP, $\cdots$ are such axes. Similarly, the axis that rotates **OC** into **OC′** must be a diameter of the great circle PN that bisects the great arc $\overset{\frown}{CC'}$ at right angles. Hence there is but one axis round which **OB**, **OC** rotate into **OB′**, **OC′**; viz., the intersection OP of the planes of these two bisecting great circles: the equator is the great circle whose plane is perpendicular to this axis, and the arcs of the rotation are the intercepts on the equator by the planes through the axis and either B, B' or C, C'. [When the two bisecting great circles coincide (as when C, C' lie on BP, $B'P$), then their plane bisects the diedral angle $BC - O - B'C'$, whose edge OP is the only axis of rotation.]

NOTE. Since $\overset{\frown}{BC}$, $\overset{\frown}{B'C'}$ may be any two positions of a marked arc on the surface of the sphere, we see that any two positions of the sphere with centre fixed determine a definite rotation of the sphere from one position to the other.

25. *A marked arc of a great circle of a rotating sphere makes a constant angle with the equator of the rotation.*

For the plane of the great arc makes a constant angle both with the axis and with the equator of the rotation.

26. *If the sphere O be given a rotation* $2\overset{\frown}{A_0C}$ *followed by a rotation* $2\overset{\frown}{CB_0}$, *the resultant rotation of the sphere is* $2\overset{\frown}{A_0B_0}$.

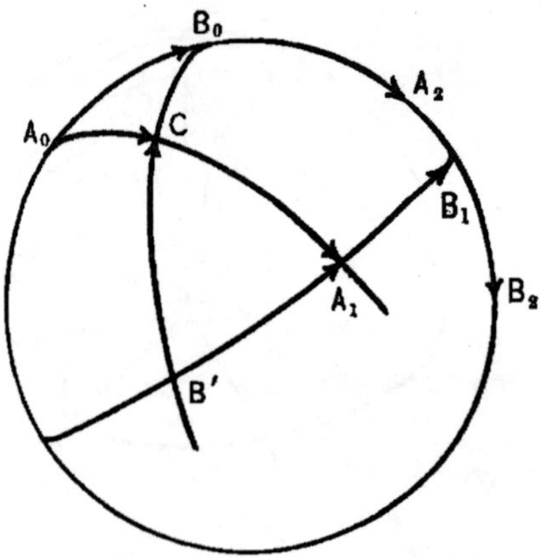

For produce the arcs $\overset{\frown}{A_0C}$, $\overset{\frown}{B_0C}$ to A_1, B' respectively, making $\overset{\frown}{CA_1} = \overset{\frown}{A_0C}$, $\overset{\frown}{B'C} = \overset{\frown}{CB_0}$. Then the spherical triangles A_0B_0C, $A_1B'C$ are equal, since the corresponding sides about the equal vertical angles at C are by construction equal. Therefore the sides $\overset{\frown}{A_0B_0}$, $\overset{\frown}{B'A_1}$ are equal in length, and the corresponding angles A_0, A_1 and B_0, B' are equal. Therefore, by Art. 25, if a marked arc $\overset{\frown}{AB}$ of the sphere coincide initially with $\overset{\frown}{A_0B_0}$, the first rotation $2\overset{\frown}{A_0C} = \overset{\frown}{A_0A_1}$ will bring $\overset{\frown}{AB}$ into the position $\overset{\frown}{A_1B_1}$ on $\overset{\frown}{B'A_1}$ produced, and the second rotation $2\overset{\frown}{CB_0} = \overset{\frown}{B'B_1}$ will bring $\overset{\frown}{AB}$ into the position $\overset{\frown}{A_2B_2}$ on $\overset{\frown}{A_0B_0}$ produced, where $\overset{\frown}{B_0A_2} = \overset{\frown}{A_0B_0}$. Hence the resultant rotation of the sphere is $2\overset{\frown}{A_0B_0} = \overset{\frown}{A_0A_2}$.

NOTE. This theorem enables one to find the resultant of any number of successive rotations, by replacing any two successive rotations by their resultant, and so on until a single resultant is found.

27. DEFINITIONS OF TURN

A step is turned when it is made to describe a *plane angle* round its initial point as centre.

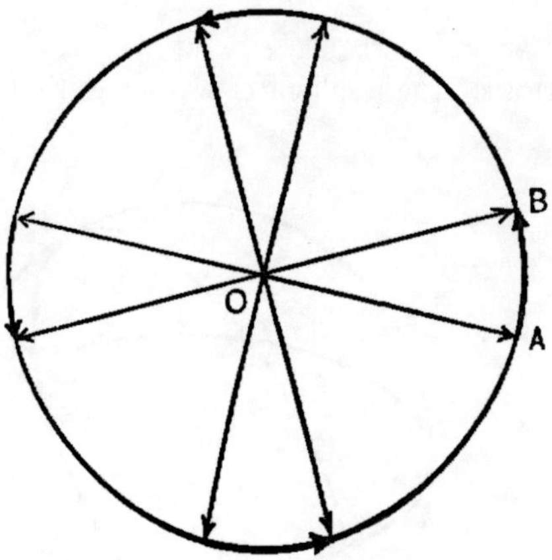

If a turn through a plane angle of given magnitude and direction in space be applied to the radii of the sphere O, it turns the great circle that is parallel to the given plane angle as a rigid circle, and does not affect the other radii of the sphere. *E.g.*, only horizontal radii can be turned through a horizontal plane angle. The circle that is so turned is the great circle of the turn.

A directed arc of the great circle of a turn from the initial position A to the final position B of a point on the great circle, and less than a semi-circumference, is the arc of the turn. When this arc is 0° or 180°, it represents a turn that brings a step back to its original position or that reverses it; and since such turns may take place in any plane with the same results, therefore such arcs may be measured on any great circle of the sphere without altering their corresponding turns.

The **axis** of a turn is that radius of the sphere O which is perpendicular to its great circle and lies on that side of the great circle from which the arc of the turn appears counter-clockwise.

28. *A turn is determined by the position into which it displaces any given step.*

For, let the radius **OA** turn into the radius **OB**. Then, the great circle $O - AB$ must be the great circle of the turn, and $\overset{\frown}{AB}$, the arc of the turn.

29. DEFINITIONS. The resultant of two successive turns $\overset{\frown}{AB}$, $\overset{\frown}{BC}$ is the turn $\overset{\frown}{AC}$.

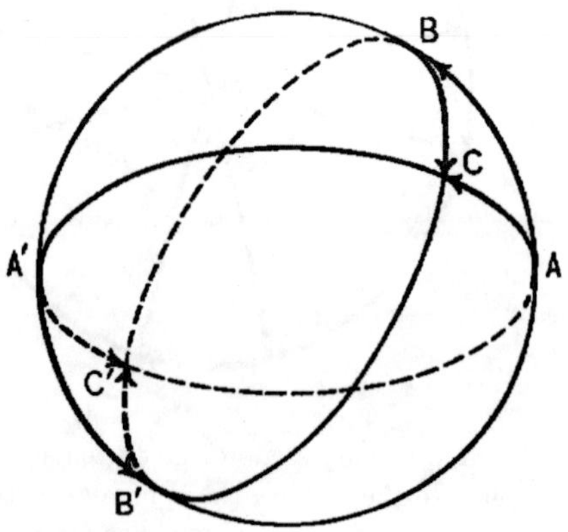

When the arc of the turns are not given with the first ending where the second begins, each arc may be moved as a rigid arc round its great circle until they do so end and begin, without altering their turning value. When the two great circles are not the same, then the common point of the two arcs must be one or the other point of intersection (B, B') of the two great circles. The figure shows that the same resultant is found from either of these points.

ARC STEPS

We may call the great arc $\frown{AB}$ the **arc step** from A to B on the surface of the sphere; and call two arc steps **equal** when they are arcs of the same great circle of the same length and direction; and call $\frown{AC}$ the **sum** of $\frown{AB}$, $\frown{BC}$ or the sum of any arc steps equal to these. The half-arc of a resultant rotation is thus the sum of the half-arcs of its components, and the arc of a resultant turn is the sum of the arcs of the components. The sum of several arcs is found by replacing any two successive arcs of the sum by their sum, and so on, until a single sum is found. An arc of 0° or 180° may be measured on any great circle without altering its value as the representative of a half-rotation, a turn, or an arc step.

30. *The resultant of two successive rotations or turns (i.e., the sum of two arc steps) is commutative only when the arcs are cocircular.*

For let the half-arcs of the rotations, or the arcs of the turns, be $\frown{AB} = \frown{BA'}$, and $\frown{C'B} = \frown{BC}$; then the sums $\frown{AB} + \frown{BC}$, $\frown{C'B} + \frown{BA'}$ in opposite orders are respectively $\frown{AC}$, $\frown{C'A'}$; and from the figure those arcs are equal when, and only when, the given arcs are cocircular.

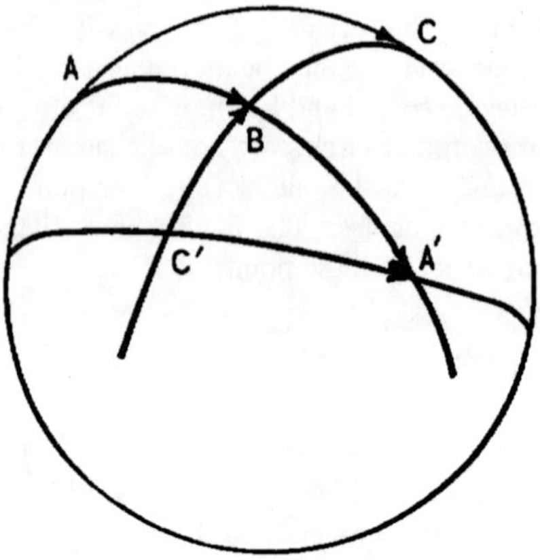

- Cor. 1. *An arc of* 0° *or* 180° *is commutative with any other arc.* For it may be taken cocircular with the other arc.
- Cor. 2. *The magnitudes of the sums of two arcs in opposite orders are equal.*

  For ABC, $A'BC'$ are equal spherical triangles by construction, and therefore $\overparen{AC}$, $\overparen{C'A'}$ are equal in length.

31. *A sum of successive arc steps is associative.*

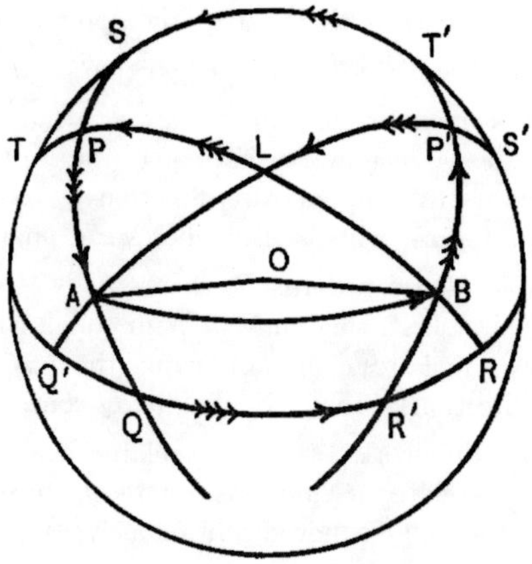

For, consider first three arcs upon the great circles LQ', $Q'R$, RL. If the arcs are such as to begin and end successively, the proof is the same as for step addition, *e.g.*, in the sum $\frown{AQ'} + \frown{Q'R} + \frown{RB} = \frown{AB}$, the first two may be replaced by their sum $\frown{AR}$, or the second and third by their sum $\frown{Q'B}$ without altering the whole sum. In the more general case when the three arcs are

$$\frown{AQ'} = \frown{S'P'}, \quad \frown{Q'Q} = \frown{R'R}, \quad \frown{RB} = \frown{PT},$$

the sum of the first two is $\frown{AQ} = \frown{SP}$, whose sum with the third is $\frown{ST}$; and the sum of the second and third is $\frown{R'B} = \frown{P'T'}$, whose sum with the first is $\frown{S'T'}$; and we must prove that $\frown{ST}$, $\frown{S'T'}$ are equal arcs of the same great circle in the same direction.

[Observe that in the construction P is determined as the intersection of QA and RB, and P' as the intersection of $Q'A$ and $R'B$.]

Let the three given arcs be the half-arcs of successive rotations of the sphere O. Then by Art. 26, the rotation $2\frown{AQ} = 2\frown{SP}$ gives the sphere the same displacement as the first and second rotations, so that $2\frown{ST}$ gives the sphere the same displacement as the three rotations. Similarly, the rotation $2\overset{\frown}{R'B} = 2\overset{\frown}{P'T}$ gives the sphere the same displacement as the second and third rotations, so that $2\overset{\frown}{S''T'}$ gives the sphere the same displacement as the three rotations. Hence $\overset{\frown}{ST}$, $\overset{\frown}{S''T'}$ are arcs of the same great circle, and either equal (and in the same direction) or supplementary (and in opposite directions), since they are half-arcs of the same rotation. This is true wherever Q may be. Suppose that Q is slightly displaced towards R; then $\overset{\frown}{ST}$, $\overset{\frown}{S''T'}$ are slightly displaced, and if equal at first, they must remain equal, since a slight change in each of two equal arcs could not change them to supplementary arcs in opposite directions.[1] Hence by moving Q continuously towards R and finding how the arcs $\overset{\frown}{ST}$, $\overset{\frown}{S'T'}$ are related when Q reaches R, we find how they are related for any position of Q, since there is no change in the relation when Q is moved continuously. But when Q is at R, it was shown above that both arcs were equal; therefore $\overset{\frown}{ST}$, $\overset{\frown}{S''T'}$ are always equal.

So, in general, for a sum of any number of successive arcs, any way of forming the sum by replacing any two successive terms by their sum and so on, must give a half-arc of the resultant of the rotations through double each of the given arcs. Hence any two such sums are either equal or opposite supplementary arcs of the same great circle; and since by continuous changes of the component arcs, they may be brought so that each begins where the preceding arc ends, in which position the two sums are equal, therefore they are always equal.

- Cor. 1. *An arc of* $0°$ *or* $180°$ *may have any position in a sum.* [Art. 30, Cor. 1.]
- Cor. 2. *The magnitude of a sum of arcs is not changed by a cyclic change in the order of its terms.*

  For $(\overset{\frown}{AB} + \overset{\frown}{CD} + \cdots) + \overset{\frown}{HK}$ and $\overset{\frown}{HK} + (\overset{\frown}{AB} + \overset{\frown}{CD} + \cdots)$ have equal magnitudes. [Art. 30, Cor. 2.]

[1] When both arcs are nearly $90°$, a slight change in each could change them from equals to supplements in the same direction.

EXAMPLES

1. Show that $2(\overset{\frown}{AB} + \overset{\frown}{BC})$ and $2\overset{\frown}{AB} + 2\overset{\frown}{BC}$ are in general unequal.

2. If (2, 30°) denote a turn of 30° counter-clockwise in the plane of the paper and a doubling, and (3, −60°) denote a turn of 60° clockwise in the plane of the paper and a trebling, express the resultant of these two compound operations (*versi-tensors*) in the same notation.

3. Find the resultant of (2, 30°), (3, 60°), (4, −120°), (1, 180°).

4. Show that either (2, −60°) or (2, 120°) taken twice have the resultant (4, −120°).

5. Would you consider the resultants of *versi-tensors* as their sums or their products, and why?

6. Let the base QR of a spherical triangle PQR slide as a rigid arc round its fixed great circle, and let the great circles QP, RP, always pass through fixed points A, B respectively. Show that if points S, T lie on the great circles QP, RP so as always to keep $\overset{\frown}{PS} = \overset{\frown}{QA}$ and $\overset{\frown}{PT} = \overset{\frown}{RB}$, then the arc $\overset{\frown}{ST}$ is an arc of fixed length and direction that slides around a fixed great circle as $\overset{\frown}{QR}$ slides round its fixed great circle. [Let P', Q', R', S', T', be given positions of P, Q, R, S, T, and use Art. 31 and figure.]

7. Show that the locus of the radius OP in Ex. 6 is an oblique circular cone of which OA, OB are two elements, and that the fixed great circles QR, ST are parallel to its circular sections. [Draw a fixed plane parallel to OQR and cutting the radii OA, OB, in the fixed points A', B', and cutting OP in the variable point P', and show that P' describes a circle in this plane through the fixed points A', B'; similarly, for a fixed plane parallel to OST.]

   NOTE.—The locus of P on the surface of the sphere is called a **spherical conic** (the intersection of a sphere about the vertex of a circular cone as centre with the surface of the cone); and the great circles QR, ST (parallel to the circular sections of the cone) are the *cyclic* great circles of the spherical conic. The above properties of a spherical conic and its cyclic great circles become properties of a plane conic and its *asymptotes* when the centre O of the sphere is taken at an indefinitely great distance.

8. State and prove Ex. 6 for a plane, and construct the locus of P.

Chapter 3

Quaternions

32. Definitions. A **quaternion** is a number that alters a step in length and direction by a given ratio of extension and a given turn. *E.g.*, in the notation of Ex. 2, II, (2, 30°), (2, −60°) are quaternions.

Two quaternions are **equal** when, and only when, their ratios of extension are equal and their turns are equal.

A **tensor** is a quaternion that extends only; *i.e.*, a tensor is an ordinary positive number. Its turn is 0° in any plane.

A **versor** or **unit** is a quaternion that turns only. *E.g.*, 1, $-1 = (1, 180°)$, $(1, 90°)$, $(1, 30°)$, are versors.

A **scalar** is a quaternion whose product lies on the same line or "scale" as the multiplicand; *i.e.*, a scalar is an ordinary positive or negative number. Its turn is 0° or 180° in any plane.

A **vector** is a quaternion that turns 90°. *E.g.*, $(2, 90°)$, $(1, -90°)$, are vectors.

33. Functions of a Quaternion q. The **tensor** of q, or briefly Tq, is its ratio of extension. *E.g.*, $T2 = 2 = T(-2) = T(2, 30°)$.

The **versor** of q (Uq) is the versor with the same arc of turn as q. *E.g.*,

$$U2 = 1, \quad U(-2) = -1, \quad U(2, 30°) = (1, 30°).$$

The **arc, angle, axis, great circle,** and **plane** of q, are respectively the *arc, angular magnitude, axis, great circle,* and *plane* of its turn.

E.g., arc$(2, 30°)$ is a counter-clockwise arc of $30°$ of unit radius in the plane of the paper, and arc$(2, -30°)$ is the same arc oppositely directed; $\angle(2, 30°) = \angle(2, -30°) = 30° = \frac{\pi}{6}$radians; axis$(2, 30°)$ is a unit length perpendicular to the plane of the paper directed towards the reader, and axis$(2, -30°)$ is the same length oppositely directed; etc.

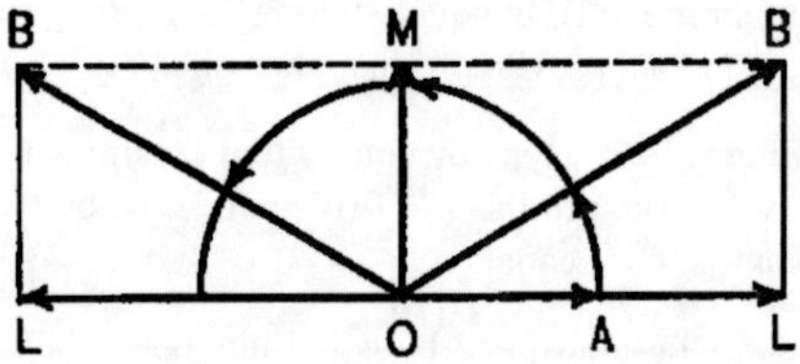

If $q\mathbf{OA} = \mathbf{OB}$, and if L be the foot of the perpendicular from B upon the line OA, then **OL**, **LB** are called *the components of q's product respectively parallel and perpendicular to the multiplicand;* also, *the projections of* **OB** *parallel and perpendicular to* **OA**.

The **scalar** of q (Sq) is the scalar whose product equals the component of q's product parallel to the multiplicand; *viz.*, $Sq \cdot \mathbf{OA} = \mathbf{OL}$.

E.g., $S(2, 30°) = \sqrt{3}, \quad S(2, 150°) = -\sqrt{3}$.

The **vector** of q (Vq) is the vector whose product equals the component of q's product perpendicular to the multiplicand; *viz.*, $Vq \cdot \mathbf{OA} = \mathbf{LB}$.

E.g., $V(2, 30°) = (1, 90°) = V(2, 150°), \quad V(2, -60°) = (\sqrt{3}, -90°)$.

The **reciprocal** of q ($1/q$ or q^{-1}) is the quaternion with reciprocal tensor and reversed turn. *E.g.*, $(2, 30°)^{-1} = (\frac{1}{2}, -30°)$.

The **conjugate** of $q(Kq)$ is the quaternion with the same tensor and reversed turn. *E.g.*,

$$K(2, 30°) = (2, -30°).$$

34. From the above diagram and the definitions of the cosine and sine of an angle, we have

(a) $$Sq = \frac{\mathbf{OL}}{\mathbf{OA}} = \frac{\overline{OL}}{OA} = \frac{OB}{OA} \cdot \frac{\overline{OL}}{OB} = Tq \cdot \cos \angle q$$

(b) $$TVq = \frac{LB}{OA} = \frac{OB}{OA} \cdot \frac{LB}{OB} = Tq \cdot \sin \angle q$$

NOTE. *Arc Vq is a quadrant on the great circle of q in the direction of arc q.*

Examples

1. If equal numbers multiply equal steps, the products are equal; and if they multiply unequal steps, the products are unequal.
2. If the products of two steps by equal numbers are equal, then the two steps are equal; and if the products of two equal steps by two numbers are equal, then the numbers are equal.
3. If several steps be multiplied by equal numbers, then any product is to its multiplicand as any other product is to its multiplicand.
4. If two steps be multiplied by reciprocal numbers, then corresponding products and multiplicands are reciprocally proportional.
5. Construct the following products, where **OA** is a unit step to the right in the plane of the paper, and determine the functions of each multiplier that are defined in Art. 33.

   (a) $2 \cdot \mathbf{OA} = \mathbf{OL}, \quad (4, 60°) \cdot \mathbf{OA} = \mathbf{OB}, \quad (4, -60°) \cdot \mathbf{OA} = \mathbf{OB}'$,
   $(2\sqrt{3}, 90°) \cdot \mathbf{OA} = \mathbf{OM}, \quad (2\sqrt{3}, -90°) \cdot \mathbf{OA} = \mathbf{OM}'$,
   $(1, 60°) \cdot \mathbf{OA} = \mathbf{OB}_1, \quad (1, -60°) \cdot \mathbf{OA} = \mathbf{OB}_1'$,
   $(1, 90°) \cdot \mathbf{OA} = \mathbf{OM}_1, \quad (1, -90°) \cdot \mathbf{OA} = \mathbf{OM}_1'$.

   (b) The same as (a) with 120° in the place of 60°.
6. Show that $SSq = Sq, \quad SVq = 0, \quad VSq = 0, \quad VVq = Vq$,
   $SKq = KSq = Sq, \quad VKq = KVq, \quad USq = \pm 1, \quad UTq = 1 = TUq$.

Multiplication

35. DEFINITION. The **product** of two or more numbers is that number whose extension and turn are the resultants of the successive extensions and turns of the factors (beginning with the right-hand factor).

    E.g., if $r\mathbf{OA} = \mathbf{OB}$, $q\mathbf{OB} = \mathbf{OC}$, $p\mathbf{OC} = \mathbf{OD}$, then we have $pqr \cdot \mathbf{OA} = pq\mathbf{OB} = p\mathbf{OC} = \mathbf{OD}$.

36. The product is, however, independent of whether a step **OA** can be found or not, such that each factor operates upon the product of the preceding factor; *i.e.*, we have by definition,

(a) $T(\cdots pqr) = \cdots Tp \cdot Tq \cdot Tr$.

(b) arc $(\cdots pqr) =$ arc $r +$ arc $q +$ arc $p + \cdots$.

37. *The product of a tensor and a versor is a number with that tensor and versor; and conversely, a number is the product of its tensor and its versor.*

For if n be a tensor, and q' a versor, then nq' turns by the factor q' and extends by the factor n, and *vice versa* for $q'n$; hence either of the products, nq', $q'n$, is a quaternion with tensor n and versor q'. Similarly,

$$q = Tq \cdot Uq = Uq \cdot Tq.$$

38. *Any successive factors of a product may be replaced by their product without altering the value of the whole product; but in general such factors can be changed in order without altering the value of the product only when those factors are cocircular.*

For replacing successive factors by their product does not alter the tensor of the whole product by Art. 36(a), nor the arc of the product by Art. 31, 36(b); but by Art. 30 the arc of the product is altered if two factors be interchanged except when those factors are cocircular.

- COR. 1. *A scalar factor may have any position in the product without altering the value of the product.* [Art. 31, Cor. 1.]
- COR. 2. *The angle of a product is not altered by a cyclic change in the order of the factors.* [Art. 31, Cor. 2.]
- COR. 3. *The scalar, and the tensor of the vector, of a product are not altered by a cyclic change in the order of the factors.* [Art. 34, *a*, *b*.]

39. *The product of two numbers with opposite turns equals the product of the tensors of the numbers; and conversely if the product of two numbers is a tensor, then the turns of the factors are opposites.* [36 *a*, *b*.]

- COR. 1. *The product of two conjugate numbers equals the square of their tensor; and if the product of two numbers with equal tensors is a tensor, then the two numbers are conjugates.*
- COR. 2. *The conjugate of a product equals the product of the conjugates of the factors in reverse order.*
  For $(pqr)(Kr \cdot Kq \cdot Kp) = (Tp)^2 \cdot (Tq)^2 \cdot (Tr)^2$ since $rKr = (Tr)^2$, may have any place in the product, and may be put first; and then $(qKq) = (Tq)^2$, may be put second, and then $(pKp) = (Tp)^2$. [Cor. 1, 38 Cor. 1.]
  Hence, $K(pqr) = Kr \cdot Kq \cdot Kp$. [Cor. 1.]
- COR. 3. *The product of two reciprocal numbers is unity; and conversely, if the product of two numbers with reciprocal tensors is unity, then the numbers are reciprocals.*
- COR. 4. *The reciprocal of a product equals the product of the reciprocals of the factors in reverse order.*
  For $(pqr)(r^{-1}q^{-1}p^{-1}) = 1$.

40. *The square of a vector is -1 times the square of its tensor; and conversely, if the square of a number is a negative scalar, then the number is a vector.* [36, *a*, *b*.]

- COR. 1. *The conjugate of a vector is the negative vector.* [39 Cor. 1.]
- COR. 2. *The conjugate of a product of two vectors is the product of the same vectors in reverse order.* [Art. 39, Cor. 2.]
- COR. 3. *The conjugate of a product of three vectors is the negative of the product of the same vectors in reverse order.* [Art. 39, Cor. 2.]

The Rotator $q()q^{-1}$

41. We may consider the ratio of two steps as determining a number, the antecedent being the product and the consequent the multiplicand of the

number; *viz.*, $\mathbf{O}B/OA$ determines the number r such that $r\mathbf{O}A = OB$. By Art. 21, equal step ratios determine equal numbers.

If the several pairs of steps that are in a given ratio r be given a rotation whose equatorial arc is 2 arc q, they are still equal ratios in their new positions and determine a new number r' that is called *the number r rotated through* 2 *arc q*. In other words, the rotation of r produces a number with the same tensor as r, and whose great circle and arc are the rotated great circle and arc of r.

42. *The number r rotated through* 2 *arc q is the number* qrq^{-1}.

For, 1st, $Tqrq^{-1} = Tq \cdot Tr(Tq)^{-1} = Tr$.

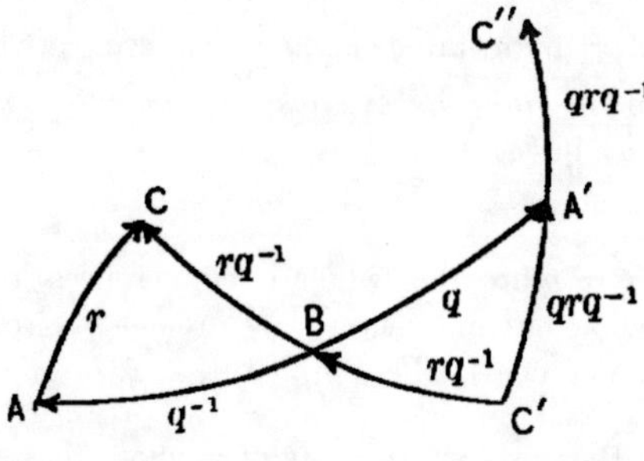

2d, let A be an intersection of the great circle of r with the great circle of q and construct

$$\overset{\frown}{AB} = \overset{\frown}{BA'} = \text{arc } q, \quad \overset{\frown}{AC} = \text{arc } r,$$

and

$$\overset{\frown}{C'B} = \overset{\frown}{BC} = \text{arc } rq^{-1};$$

then

$$\overset{\frown}{C'A'} = \overset{\frown}{A'C''} = \text{arc } qrq^{-1}.$$

But by construction, the spherical triangles ABC, $A'BC'$ are equal, and therefore $\overset{\frown}{AC}$ and $\overset{\frown}{C'A'}(=\overset{\frown}{A'C''})$ are arcs of equal length, and the corresponding angles at A, A' are equal. Hence, when arc $r(=\overset{\frown}{AC})$ is rotated through 2 arc $q(=\overset{\frown}{AA'})$, it becomes arc $qrq^{-1}(=\overset{\frown}{A'C''})$.

Powers and Roots

43. An integral power, $q^n = q \cdot q \cdot q \cdots$ *to n factors*, is determined by the equations,

(a) $T \cdot q^n = Tq \cdot Tq \cdot Tq \cdots = (Tq)^n$.

(b) arc $q^n =$ arc $q+$arc $q+$arc $q \cdots = n$ arc $q\pm$(whole circumferences). To find $q^{\frac{1}{n}}$, *the number whose nth power is q*, we have, by replacing q by $q^{\frac{1}{n}}$ in (a), (b),

(c) $Tq = (T \cdot q^{\frac{1}{n}})^n$ or $T \cdot q^{\frac{1}{n}} = (Tq)^{\frac{1}{n}}$

(d) Arc $q = n$ arc $q^{\frac{1}{n}}\pm$ whole circumferences, or, arc $q^{\frac{1}{n}} = \frac{1}{n}$ (arc $q\pm$ whole circumferences) $= \frac{1}{n}$ arc $q + \frac{m}{n}$ circumferences $\pm$ whole circumferences), where $m = 0, 1, 2, 3, \cdots n-1$, successively.

There are therefore n nth roots of q whose tensors are all equal and whose arcs lie on the great circle of q.

When the base is a scalar, its great circle may be any great circle, so that there are an infinite number of quaternion nth roots of a scalar. On this account, the roots as well as the powers of a scalar are **limited to scalars.** By ordinary algebra, there are n such nth roots, real and imaginary. There are also imaginary nth roots of q besides the n real roots found above; *i.e.*, roots of the form $a + b\sqrt{-1}$, where a, b are real quaternions.

Representation of Vectors

44. Bold-face letters will be used as symbols of vectors only. In particular, **i**, **j**, **k** will denote *unit* vectors whose axes are respectively a unit length

east, a unit length north, and a unit length up. More generally we shall use the step **AB** to denote the vector whose axis is a unit length in the direction of **AB**, and whose tensor is the numerical length of **AB** (= AB: *unit* length).

This use of a step **AB** as the symbol of a vector is analogous to the use of AB to represent a tensor (AB: *unit* length), or of $\overline{AB}$ to represent a positive or negative scalar, according as it is measured in or against the direction of its axis of measurement. In none of these cases is the concrete quantity an absolute number; *i.e.*, the value of the number that it represents varies with the assumed unit of length. When desirable, we distinguish between the vector **OA** and the step **OA** by enclosing the vector in a parenthesis.

45. *If* $q(\mathbf{OA}) = (\mathbf{OB})$, *then* $q \cdot \mathbf{OA} = \mathbf{OB}$, *and conversely.*

The tensor of q in either equation is $OB : OA$. It is therefore only necessary to show that the arc of q in one equation equals the arc of q in the other equation in order to identify the two numbers that are determined by these two equations as one and the same number.

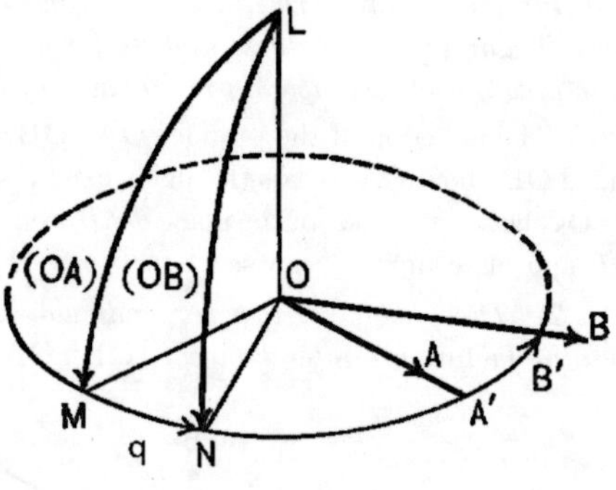

Draw the sphere of unit radius and centre O, cutting **OA**, **OB** in A', B'; then $\overset{\frown}{A'B'}$ is the arc of q in the second equation. Draw the radius OL perpendicular to the plane $OA'B'$ on the counter-clockwise side of $\overset{\frown}{A'B'}$, and draw counter-clockwise round OA', OB' as axes the quadrants $\overset{\frown}{LM}$, $\overset{\frown}{LN}$ respectively; then these are the arcs of (**OA**), (**OB**)

respectively, and since $\overset{\frown}{LM} + \overset{\frown}{MN} = \overset{\frown}{LN}$, therefore $\overset{\frown}{MN}$ is the arc of q in the first equation. But since $\overset{\frown}{LM}$, $\overset{\frown}{LN}$ are quadrants, therefore the plane OMN is perpendicular to OL, and must therefore coincide with the plane $OA'B'$, which is by construction also perpendicular to OL. Hence $\overset{\frown}{MN}$ lies on the great circle of $\overset{\frown}{A'B'}$, and by the construction of the figure, it must, when advanced 90° on that great circle, coincide with $\overset{\frown}{A'B'}$. Hence the theorem.

NOTE. This theorem shows that a number extends and turns vectors into vectors in the same way that it extends and turns steps into steps. Moreover, when the vector is not perpendicular to the axis of the multiplier, there is no resulting vector, since in the case of the corresponding step there is no resulting step. In the case of a vector multiplicand, that is oblique to the axis of q, the product is an actual quaternion that is not a vector, while in the case of the corresponding step multiplicand the product belongs to that class of products in which the multiplicand does not admit of the operation of the multiplier, as in $\sqrt{2}$ *universities, -2 countries, etc.*

- COR. 1. *The product of two vectors is a vector when, and only when, the factors are perpendicular to each other; the product is perpendicular to both factors; and its length (its tensor) is equal to the area of the rectangle on the lengths of the factors.*

  NOTE. The direction of the product $\mathbf{OA} \cdot \mathbf{OB} = \mathbf{OC}$ is obtained by turning **OB** about **OA** as axis through a counter-clockwise right angle; thus **OC** lies on that side of the plane OAB from which the right angle AOB appears counter-clockwise.
- COR. 2. *The product of two perpendicular vectors changes sign when the factors are interchanged.* ($\mathbf{OB} \cdot \mathbf{OA} = \mathbf{OC'} = -\mathbf{OC}$.)

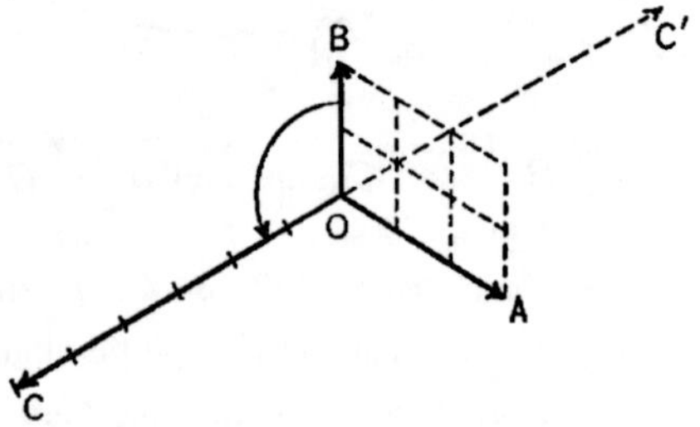

- COR. 3. *The condition that α is perpendicular to β is that $\alpha\beta =$ vector, or $S\alpha\beta = 0$.*

46. *If* **AB**, **CD** *are parallel, then* $\mathbf{AB}\cdot\mathbf{CD} = \mathbf{CD}\cdot\mathbf{AB} = -\overline{AB}\cdot\overline{CD}$, *a scalar; and conversely, the product of two vectors is a scalar only when they are parallel.*

Since the axes of the vectors **AB**, **CD** are parallel, therefore their product is commutative. When the vectors are in the same direction, then each turns 90° in the same direction, the resultant turn is 180°, and the product is negative; and when the vectors are in opposite direction, their turns are in opposite directions, the resultant turn is 0°, and the product is positive. This is just the opposite of the product of the corresponding scalars $\overline{AB}$, $\overline{CD}$, which is positive when the scalars are in the same direction (or both of the same sign), and negative when the scalars are in opposite directions; *i.e.*, $\mathbf{AB}\cdot\mathbf{CD} = -\overline{AB}\cdot\overline{CD}$.

Conversely, the product **AB**, **CD** can be a scalar only when the resultant of their two turns of 90° each is a turn of 0° or 180°; *i.e.*, only when the turns are cocircular, and therefore their axes parallel.

- COR. *The condition that α is parallel to β is $\alpha\beta =$ scalar, or $V\alpha\beta = 0$.*

Examples

1. Prove by diagram that $(pq)^2$ and p^2q^2 are in general unequal.
2. Find the 2d, 3d, 4th, 5th, 6th powers of $(2, 90°)$, $(2, -60°)$.
3. Find the square roots and cube roots of $(4, 30°)$, $(8, -120°)$.
   (a) Find the values of $[(2, 50°)^6]^{\frac{1}{3}}$, $[(2, 50°)^{\frac{1}{3}}]^6$, and $(2, 50°)^{\frac{6}{3}}$.
4. What numbers are represented by 2 *feet*, 2 *feet east*, the unit of length being a foot, a yard, an inch?
5. Show that $\mathbf{i^2 = j^2 = k^2 = ijk = -1}$; $\mathbf{jk = i = -kj}$; $\mathbf{ki = j = -ik}$; $\mathbf{ij = k = -ji}$.

6. Let $e^{(\mathbf{AB})}$ denote the versor that turns counter-clockwise round the axis AB through an arc that is formed by bending the length AB into an arc of unit radius. Show that if facing the west, and holding the paper in a north and south vertical plane, then $e^{\mathbf{i}}$, $e^{2\mathbf{i}}$, $\cdots e^{-\mathbf{i}}$, $e^{-2\mathbf{i}}$, turn respectively 1, 2, $\cdots$ radians counter-clockwise, and 1, 2, $\cdots$ radians clockwise in the plane of the paper. Also show that $e^{\pm\frac{\pi}{2}\mathbf{i}} = \pm\mathbf{i}$, $e^{\pm\pi\mathbf{i}} = -1$ $e^{2n\pi\mathbf{i}} = 1$, where n is any integer.

7. Show by diagram that $Se^{\theta\mathbf{i}} = \cos\theta$, $Ve^{\theta\mathbf{i}} = i\sin\theta$, where θ is any positive or negative number and the unit of angle is a radian.

8. Show that if **OA** rotate into **OB** through 2 arc q, then $(\mathbf{OB}) = q(\mathbf{OA})q^{-1}$.

9. Show that if α be a vector in the plane of q, then $Kq = \alpha q\alpha^{-1} = \alpha^{-1}q\alpha$.

10. Show that pq rotates into qp, and determine two such rotations.

11. Show that $SKq = Sq$, $VKq = -Vq$.

12. Show that $K\alpha\beta = \beta\alpha$, $S\alpha\beta = S\beta\alpha$, $V\alpha\beta = -V\beta\alpha$.

13. Show that $K\alpha\beta\gamma = -\gamma\beta\alpha$; $V\alpha\beta\gamma = V\gamma\beta\alpha$; $S\alpha\beta\gamma = S\beta\gamma\alpha = S\gamma\alpha\beta = -S\gamma\beta\alpha = -S\beta\alpha\gamma = -S\alpha\gamma\beta$. ($a$) Determine the conjugate of a product of n vectors.

14. Prove by diagram that $Kpq = Kq \cdot Kp$.

Addition

47. DEFINITION. The sum $(p+q)$ is the number determined by the condition that its product is the sum of the products of p and q.

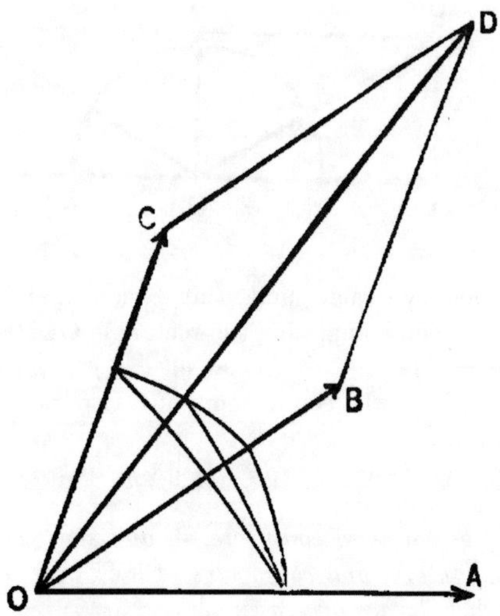

Thus let **OA** be any step that is multiplied by both p and q, and let $p\mathbf{OA} = \mathbf{OB}$, $q\mathbf{OA} = \mathbf{OC}$, and $\mathbf{OB} + \mathbf{OC} = \mathbf{OD}$, then $(p + q)\mathbf{OA} = \mathbf{OD}$. It is obvious that any change in **OA** alters **OB**, **OC**, **OD**, proportionally, so that the value of the sum $p + q (= \mathbf{OD} : \mathbf{OA})$ is the same for all possible values of **OA**.

Similarly, any quaternion, r, may be added to the sum $p + q$, giving the sum $(p+q)+r$; and we may form other sums such as $p+(q+r)$, $(q+r)+p$, etc. It will be shown later that all such sums of the same numbers are equal, or that quaternion addition is *associative* and *commutative.*

48. *The sum of a scalar and a vector is a quaternion with that scalar and that vector, and conversely, a quaternion is the sum of its scalar and its vector.*

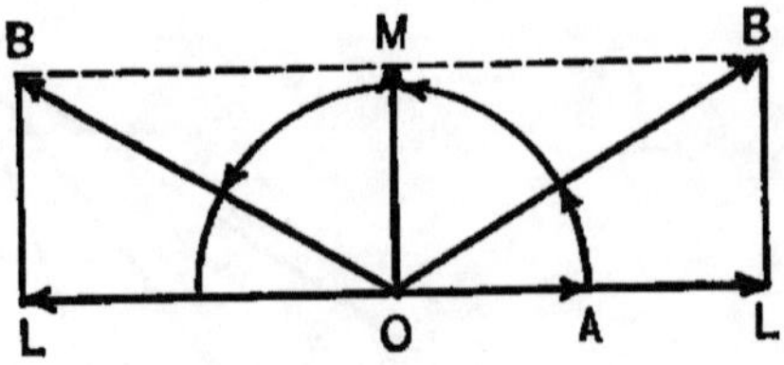

For let w be any scalar, and ρ any vector, and let $w\mathbf{OA} = \mathbf{OL}$, $\rho\mathbf{OA} = \mathbf{OM}$, then completing the rectangle **OLBM**, we have $(w + \rho)\mathbf{OA} = \mathbf{OB}$, and the scalar of $w + \rho$ is w, and its vector is ρ, since **OL**, **OM** are the components of **OB** parallel and perpendicular to **OA**. Similarly,

$$q = Sq + Vq.$$

49. *The scalar, vector, and conjugate, of any sum equals the like sum of the scalars, vectors, and conjugates of the terms of the sum.* [*I.e.*, S, V, K, *are distributive over a sum.*]

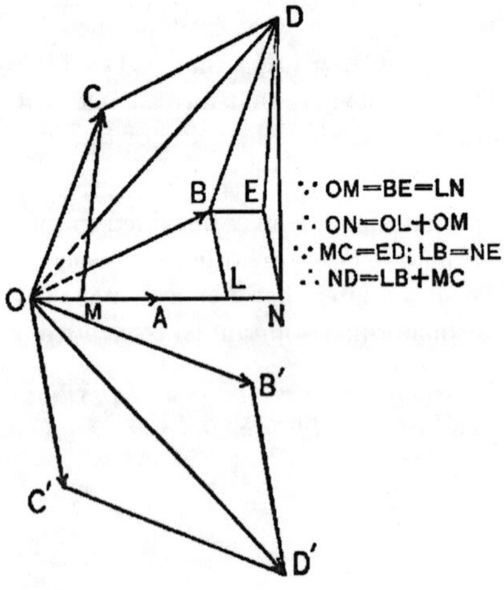

For let

$$p\mathbf{OA} = \mathbf{OB}, \quad q\mathbf{OA} = \mathbf{OC},$$
$$(p+q)\mathbf{OA} = \mathbf{OB} + \mathbf{OC} = \mathbf{OD}.$$

Then the components of **OD** parallel and perpendicular to **OA** are, by the figure, the sums of the like components of **OB**, **OC**; *i.e.*, $S(p+q)\cdot\mathbf{OA} = Sp\cdot\mathbf{OA} + Sq\cdot\mathbf{OA}$, or $S(p+q) = Sp + Sq$; and $V(p+q)\cdot\mathbf{OA} = Vp\cdot\mathbf{OA} + Vq\cdot\mathbf{OA}$, or $V(p+q) = Vp + Vq$.

Also, if $OB'D'C'$ be the parallelogram that is symmetric to the parallelogram $OBDC$ with reference to OA as axis of symmetry, then $Kp\cdot\mathbf{OA} = \mathbf{OB}'$, $Kq\cdot\mathbf{OA} = \mathbf{OC}'$, and $K(p+q)\cdot\mathbf{OA} = \mathbf{OD}'$, and since $\mathbf{OB}' + \mathbf{OC}' = \mathbf{OD}'$, therefore $K(p+q) = Kp + Kq$.

These results extend to any given sum; *e.g.*, $V[(p+q)+r] = V(p+q) + Vr = (Vp + Vq) + Vr$, etc.

50. *If* $(\mathbf{OA}) + (\mathbf{OB}) = (\mathbf{OC})$, *then* $\mathbf{OA} + \mathbf{OB} = \mathbf{OC}$, *and conversely.*

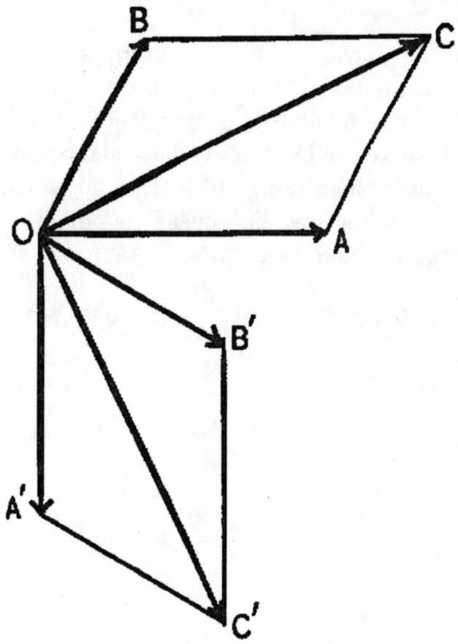

For erect a pin OD of unit length perpendicular to the plane of the angle AOB on its counter-clockwise side; and turn AOB round OD as axis through a clockwise right angle as seen from D into the position $A'OB'$. Then since $(\mathbf{O}A)$ is the vector that turns through a counter-clockwise right angle round OA as axis, and extends unit length into $OA = OA'$, therefore $(OA)OD = OA'$, and similarly $(OB)OD = OB'$, and therefore $(OC)OD = OA' + OB' = OC'$, where $OA'C'B'$ is a parallelogram. Hence the step $\mathbf{O}C$ of proper length and direction to give the tensor and axis of the vector (OC) must be the diagonal of the parallelogram on $\mathbf{O}A$, $\mathbf{O}B$ as sides; and therefore $\mathbf{O}A + OB = OC$. Conversely, if $\mathbf{O}A + OB = OC$, then turning the parallelogram $OACB$ into the position $OA'C'B'$, we have, since $\mathbf{O}A' + OB' = OC'$, that $(OA) + (OB) = (OC)$.

- COR. 1. *Vectors add in the same way as their corresponding steps, and all the laws of addition and resolution of steps extend at once to vectors.*
- COR. 2. *A sum of quaternions is associative and commutative.*

For since by Cor. 1 a sum of vectors is independent of the way in which its terms are added, and since we know that a sum of scalars (i.e., ordinary numbers) is independent of the way in which its terms are added, therefore by Art. 49 the scalar and the vector of a sum are independent of the way in which the sum is added. Hence the sum is independent of the way in which it is added, since it is equal to the sum of its scalar and its vector.

51. LEMMA. *If p, q be any quaternions, then $(1 + p)q = q + pq$.*

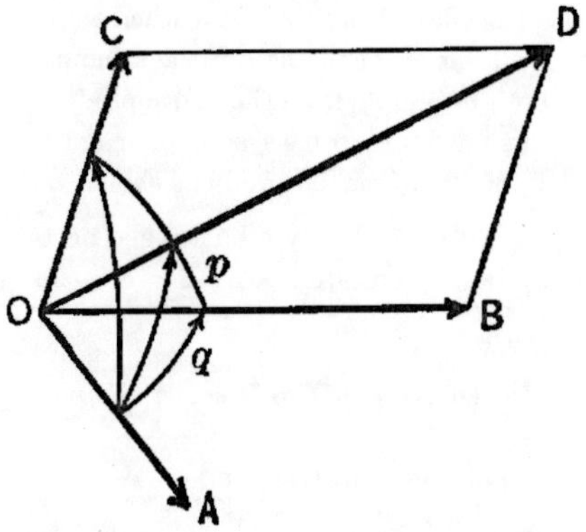

For take **OB** in the intersection of the planes of p, q and draw **OA**, **OC** such that $q \cdot \mathbf{OA} = \mathbf{OB}$, $p\mathbf{OB} = \mathbf{OC}$; then $(1+p)q \cdot \mathbf{OA} = (1+p)\mathbf{OB} = \mathbf{OB} + \mathbf{OC} = q\mathbf{OA} + pq\mathbf{OA}$. Hence,

$$(1+p)q = q + pq.$$

52. *If* p, q, r *be any quaternions, then* $(p+q)r = pr + qr$.

For we have, $(1 + qp^{-1})p \cdot r = (1 + qp^{-1}) \cdot pr$, and expanding each member by the preceding lemma, we have, $(p+q)r = pr + qr$.

This result extends to any sum; *e.g.*,

$$(p+q+r+s)t = [(p+q)+(r+s)]t = (p+q)t+(r+s)t = pt+qt+rt+st.$$

- Cor. 1. $r(p+q) = rp + rq$.
  For let p', q', r' be the conjugates of p, q, r. Then from $(p'+q')r' = p'r' + q'r'$, we have, by taking the conjugates of each member, $r(p+q) = rp + rq$. [Art. 39, Cor. 2; Art. 49.]
- Cor. 2. *A product of sums equals the sum of all partial products that may be formed from the given product by multiplying together, in the order in which they stand, a term from each factor of the product.*
  E.g., $(p+q)(r+s) = pr + ps + qr + qs$.

NOTE.—This rule should be used even when the factors are commutative, as it prevents all danger of taking out the same partial product twice; *e.g.*, from taking both pr and rp from the above product. To be sure that all the partial products are found, some system of arrangement should be adopted; also the total number of partial products should be determined.

E.g., $(p+q)(p+q)(p+q)$ may be arranged according to the degrees of the terms in p, and there are $2 \times 2 \times 2 = 8$ terms. This product is then easily seen to be

$$p^3 + (p^2q + pqp + qp^2) + (pq^2 + qpq + q^2p) + q^3,$$

when p, q are not commutative, and

$$p^3 + 3p^2q + 3pq^2 + q^3,$$

when p, q are commutative.

Formulas. For Exercise and Reference

53. (a) $q = Tq \cdot Uq = Uq \cdot Tq$.
    (b) $q = Sq + Vq$; $Kq = Sq - Vq$.
    (c) $Sq = Tq \cos \angle q, = r\cos\theta$, say; $TVq = Tq \sin \angle q = r \sin\theta$.
    (d) $Vq = TVq \cdot UVq = r \sin\theta\epsilon$ where $\epsilon = UVq$.
    (e) $q = r(\cos\theta + \epsilon \cdot \sin\theta) = re^{\theta\epsilon}$, $Kq = r(\cos\theta - \epsilon \sin \sin\theta) = re^{-\theta\epsilon}$
    (f) $e^{\theta\epsilon} \cdot e^{\theta'\epsilon} = e^{(\theta+\theta')\epsilon}$.
    (g) $Sq = \frac{1}{2}(q + Kq)$; $Vq = \frac{1}{2}(q - Kq)$.
    (h) $Tq^2 = qKq = Kq \cdot q = (Sq)^2 - (Vq)^2 = (Sq)^2 + (TVq)^2$.
    (i) $q^{-1} = Kq/Tq^2$.

    As a further exercise find the T, U, S, V, K *of the* T, U, S, V, K *of* q, *in terms of* r, θ, ϵ.

54. (a) $T(\cdots pqr) = \cdots Tp \cdot Tq \cdot Tr$.
    (b) $U(\cdots pqr) = \cdots Up \cdot Uq \cdot Ur$.

(c) $\angle(\cdots pqr) = \angle(r \cdots pq) = \angle(qr \cdots p)$, etc.

(d) $S(\cdots pqr) = S(r \cdots pq) = S(qr \cdots p)$, etc.

(e) $TV(\cdots pqr) = TV(r \cdots pq) = TV(qr \cdots p)$, etc.

(f) arc $(\cdots pqr) =$ arc $r +$ arc $q +$ arc $p + \cdots$.

(g) $(\cdots pqr)^{-1} = r^{-1}q^{-1}p^{-1}\cdots$.

(h) $K(\cdots pqr) = Kr \cdot Kq \cdot Kp \cdots$.

(i) $S(xp+yq+zr) = xSp+ySq+zSr$, [$x$, y, z, *scalars*] *and similarly for* V *or* K *instead of* S.

55. (a) $K\alpha = -\alpha$; $T\alpha^2 = -\alpha^2$; $S\alpha = 0$; $V\alpha = \alpha$.

(b) $K\alpha\beta = \beta\alpha$; $S\alpha\beta = S\beta\alpha$; $V\alpha\beta = -V\beta\alpha$.

(c) $\alpha\beta + \beta\alpha = 2S\alpha\beta$, $\alpha\beta - \beta\alpha = 2V\alpha\beta$.

(d) $(\alpha \pm \beta)^2 = \alpha^2 \pm 2S\alpha\beta + \beta^2$.

(e) $V(x\alpha + y\beta)(x'\alpha + y'\beta) = (xy' - x'y)V\alpha\beta = \begin{vmatrix} x & y \\ x' & y' \end{vmatrix} V\alpha\beta$, say. [$x$, y, x', y', scalars.]

(f) $V(x\alpha + y\beta + z\gamma)(x'\alpha + y'\beta + z'\gamma) = \begin{vmatrix} y & z \\ y' & z' \end{vmatrix} V\beta\gamma + \begin{vmatrix} z & x \\ z' & x' \end{vmatrix} V\gamma\alpha + \begin{vmatrix} x & y \\ x' & y' \end{vmatrix} V\alpha\beta$. [$x$, y, z, x', y', z', scalars.]

56. (a) $K\alpha\beta\gamma = -\gamma\beta\alpha$.

(b) $\alpha\beta\gamma - \gamma\beta\alpha = 2S\alpha\beta\gamma = -2S\gamma\beta\alpha$.

Hence the scalars of the six products of α, β, γ are equal to one of two negative numbers according to the cyclic order of the product; and an interchange in two factors (which changes the cyclic order) changes the sign of the scalar of the product. When two of the three factors are equal, the scalar of their product must therefore be zero, since an interchange of the equal factors changes the sign without changing the value.

(c) $S \cdot (x\alpha + y\beta + z\gamma)(x'\alpha + y'\beta + z'\gamma)(x''\alpha + y''\beta + z''\gamma)$

$$= \left\{ x \begin{vmatrix} y' & z' \\ y'' & z'' \end{vmatrix} + y \begin{vmatrix} z' & x' \\ z'' & x'' \end{vmatrix} + z \begin{vmatrix} x' & y' \\ x'' & y'' \end{vmatrix} \right\} S\alpha\beta\gamma$$

$$= \begin{vmatrix} x & y & z \\ x' & y' & z' \\ x'' & y'' & z'' \end{vmatrix} S\alpha\beta\gamma, \text{ say. } [x, y, z, \text{etc., scalars.}]$$

(d) $S\alpha\beta\gamma = S\alpha V\beta\gamma = S\beta V\gamma\alpha = S\gamma V\alpha\beta$.

[Replace $\beta\gamma$ by $S\beta\gamma + V\beta\gamma$, expand by 54 (i), and note that $S \cdot \alpha S\beta\gamma = 0$.]

(e) $\alpha\beta\gamma + \gamma\beta\alpha = 2V\alpha\beta\gamma = 2V\gamma\beta\alpha$.

NOTE.—Insert between the two terms of the first member of (e), the null term $(\alpha\gamma\beta - \alpha\gamma\beta - \gamma\alpha\beta + \gamma\alpha\beta)$, and it becomes $\alpha(\beta\gamma + \gamma\beta) - (\alpha\gamma + \gamma\alpha)\beta + \gamma(\alpha\beta + \beta\alpha)$. Hence, using (55 c), we have (f).

(f) $V\alpha\beta\gamma = \alpha S\beta\gamma - \beta S\gamma\alpha + \gamma S\alpha\beta$.

Transpose the first term of the second member of (f) to the first member, noting that $\alpha S\beta\gamma = V \cdot \alpha S\beta\gamma$, and $\beta\gamma - S\beta\gamma = V\beta\gamma$, and we have

(g) $V\alpha V\beta\gamma = -\beta S\gamma\alpha + \gamma S\alpha\beta$;

(g') $V \cdot (V\beta\gamma)\alpha = \beta S\gamma\alpha - \gamma S\alpha\beta$.

(h)

$$[(g), (d)] \qquad V \cdot (V\alpha\beta)V\gamma\delta = -\gamma S\alpha\beta\delta + \delta S\alpha\beta\gamma$$
$$[(g'), (d)] \qquad = \alpha S\beta\gamma\delta - \beta S\alpha\gamma\delta.$$

(i)

$$[(h)] \qquad \delta S\alpha\beta\gamma = \alpha S\beta\gamma\delta + \beta S\gamma\alpha\delta + \gamma S\alpha\beta\delta.$$

Replace α, β, γ, by $V\beta\gamma, V\gamma\alpha, V\gamma\beta$, noting that $V \cdot (V\gamma\alpha \cdot V\alpha\beta) = -\alpha S\alpha\beta\gamma$, etc., and that $S(V\beta\gamma \cdot V\gamma\alpha \cdot V\alpha\beta) = -(S\alpha\beta\gamma)^2$, and we have

(j) $\delta S\alpha\beta\gamma = V\beta\gamma S\alpha\delta + V\gamma\alpha S\beta\delta + V\alpha\beta S\gamma\delta$.

NOTE.—(i), (j) may be obtained directly by putting $\delta = x\alpha + y\beta + z\gamma$ or $xV\beta\gamma + yV\gamma\alpha + zV\alpha\beta$, and finding x, y, z, by multiplying in the first case by $\beta\gamma$, $\gamma\alpha$, $\alpha\beta$, and in the second case by α, β, γ, and taking the scalars of the several products.

57. (a) $\mathbf{i}^2 = \mathbf{j}^2 = \mathbf{k}^2 = \mathbf{ijk} = -1$; $\mathbf{jk} = \mathbf{i} = -\mathbf{kj}$; $\mathbf{ki} = \mathbf{j} = -\mathbf{ik}$, $\mathbf{ij} = \mathbf{k} = -\mathbf{ji}$.

(b) $\rho = \mathbf{i}S\mathbf{i}\rho - \mathbf{j}S\mathbf{j}\rho - \mathbf{k}S\mathbf{k}\rho$. [56 (i) or (j) or directly as in note.]

Let $\rho = x\mathbf{i} + y\mathbf{j} + \mathbf{z}k$, $\rho' = x'\mathbf{i} + y'\mathbf{j} + z'\mathbf{k}$, etc. [$x$, y, z, etc. scalars.] Then, prove by direct multiplication,

(c) $-\rho^2 = x^2 + y^2 + z^2 = T\rho^2$.

(d) $-S\rho\rho' = xx' + yy' + zz' = -s\rho'\rho$.

(e) $v\rho\rho' = \begin{vmatrix} y & z \\ y' & z' \end{vmatrix} \mathbf{i} + \begin{vmatrix} z & x \\ z' & x' \end{vmatrix} \mathbf{j} + \begin{vmatrix} x & y \\ x' & y' \end{vmatrix} \mathbf{k} = -V\rho'\rho$.

(f) $-S\rho\rho'\rho'' = \begin{vmatrix} x & y & z \\ x' & y' & z' \\ x'' & y'' & z'' \end{vmatrix} = -S\rho V\rho'\rho''$.

Geometric Theorems

58. *The angle of $\alpha\beta$ equals the supplement of the angle θ between α, β.*

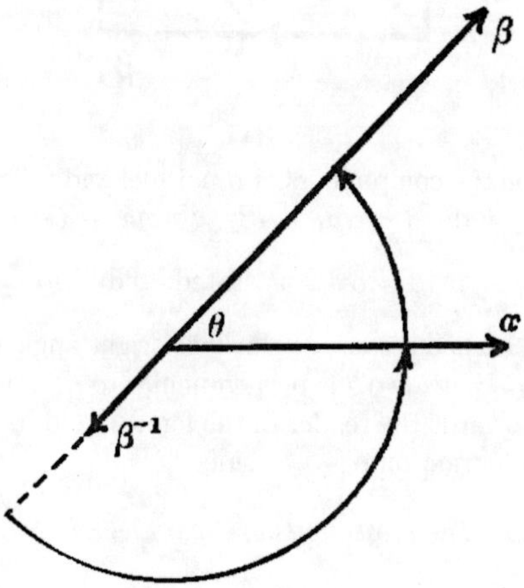

For, since $\alpha\beta \cdot \beta^{-1} = \alpha$, therefore $\alpha\beta$ turns through the angle from β^{-1} to α, which is the supplement of the angle θ from α to β.

- Cor. $S\alpha\beta = -T\alpha\beta\cos\theta$, $TV\alpha\beta = T\alpha\beta\sin\theta$. [$Sq = Tq\cos\angle q$, etc.]

59. *The scalar of $\alpha\beta$ equals the product of α and the projection of β upon it; the vector of $\alpha\beta$ equals the product of α and the projection of β perpendicular to it, and $V\alpha\beta$ is a vector perpendicular to α, β on their counter-clockwise side whose length equals the area of the parallelogram on α, β as sides.*

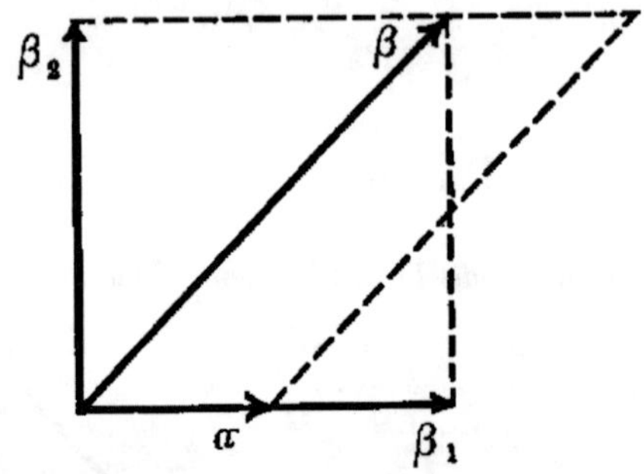

Let β_1, β_2 be the components of β parallel and perpendicular to α, then $\beta = \beta_1 + \beta_2$ and $\alpha\beta = \alpha\beta_1 + \alpha\beta_2 = \textit{scalar} + \textit{vector}$. Hence

$$S\alpha\beta = \alpha\beta_1, \text{ as stated; and } V\alpha\beta = \alpha\beta_2,$$

which is β_2 turned a counter-clockwise right angle round α and lengthened by $T\alpha$. Hence $V\alpha\beta$ is perpendicular to α, β on their counterclockwise side (towards the reader in the figure), and its length is $T\alpha \cdot T\beta_2 =$ area parallelogram on α, β, as sides.[1]

- COR. 1. *The projections of β parallel and perpendicular to α equal $\alpha^{-1}S\alpha\beta$ and $\alpha^{-1}V\alpha\beta$.*
- COR. 2. *The scalar measure of the projection of β upon α is $\frac{-S\alpha\beta}{T\alpha}$, and the tensor measure of the projection of β perpendicular to α is $\frac{TV\alpha\beta}{T\alpha}$.* [Also from 58, Cor.]

[1]The parallelogram on α, β may be considered as bounded by the path of a point that receives the displacement α, then the displacement β, then the displacement $-\alpha$, then the displacement $-\beta$. This area is therefore bounded *counter-clockwise* round $V\alpha\beta$ as axis; and $V\alpha\beta$ may therefore be called the vector measure of the area of this directed parallelogram or of any parallel plane area of the same magnitude and direction of boundary.

- COR. 3. *If θ be the angle between α, β, then* $\cos\theta = -\frac{S\alpha\beta}{T\alpha\beta}$, $\sin\theta = \frac{TV\alpha\beta}{T\alpha\beta}$. [Also from 58, Cor.]

60. *The volume of a parallelepiped on α, β, γ as edges is $-S\alpha\beta\gamma$ (the volume being positive or negative according as α lies on the counter-clockwise or clockwise side of β, γ).*

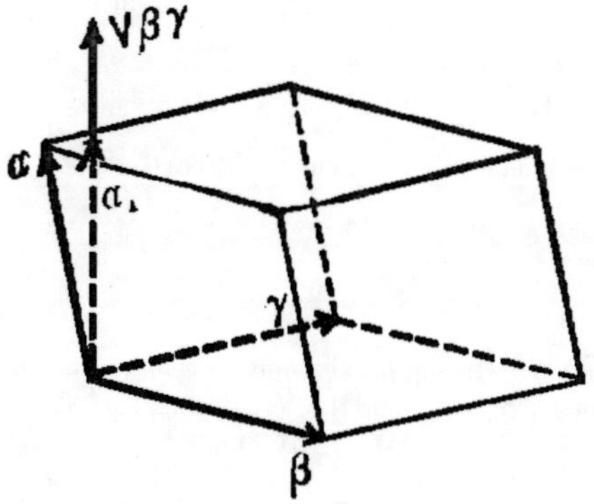

For let α_1 be the projection of α upon $V\beta\gamma$; then taking the face β, γ of the parallelepiped as the base, we have by Art. 59 that $TV\beta\gamma$ is the area of the base; also $T\alpha_1$ is the altitude. Hence *numerical volume*

[Art. 46.] $\qquad = T\alpha_1 \cdot TV\beta\gamma = \mp\alpha_1 V\beta\gamma$

[59, 56, *d.*] $\qquad = \mp S\alpha V\beta\gamma = \mp S\alpha\beta\gamma.$

The upper or lower sign must be taken according as α_1, $V\beta\gamma$ are in the same or opposite directions. This numerical result must be multiplied by -1 when α lies on the clockwise side of β, γ; i.e., when α_1, $V\beta\gamma$ are opposites (since $V\beta\gamma$ lies on the counter-clockwise side of β, γ). Hence—$S\alpha\beta\gamma$ is the required algebraic volume.

- COR. *The condition that α, β, γ are coplanar vectors is that $S\alpha\beta\gamma = 0$ (or $\alpha\beta\gamma$ = a vector).*

Examples

1. Expand $(p+q+r)^2$, $(\alpha+\beta+\gamma)^2$, $(p+q)(p-q)$, $(p-q)(p+q)$, $(p+q)K(p+q)$. Show that $T(p+q)^2 = Tp^2 + 2SpKq + Tq^2$.

2. Solve $q^2+4\mathbf{k}q-8 = 0$ for q. [q must be cocircular with $\mathbf{k}$. Hence $q = -2\mathbf{k}\pm 2$ are the real solutions.]

3. Find the tensor, versor, scalar, vector, and angle of each of the numbers: 2, -3, $3\mathbf{i}$, $2+3\mathbf{i}$, $\mathbf{i}+\mathbf{j}$, $3\mathbf{i}+4\mathbf{j}$, $5e^{\frac{\pi}{3}\mathbf{i}}$, $(2\mathbf{i}+3\mathbf{j}+6\mathbf{k})^2$.

4. Show that the three quaternion cube roots of -1, with horizontal great circle, are -1, $\frac{1}{2}\pm\frac{1}{2}\sqrt{3}\mathbf{k}$.

5. Show geometrically that $e^{\theta\epsilon}+e^{-\theta\epsilon} = 2\cos\theta$, $\epsilon^{\theta\epsilon}-e^{-\theta\epsilon} = 2\epsilon\sin\theta$.

6. The numbers e^α and
$$1+\alpha+\frac{\alpha^2}{2!}+\frac{\alpha^3}{3!}+\frac{\alpha^4}{4!}+\cdots$$
are equal. Verify this approximately by geometric construction when $T\alpha = 1$, and when $T\alpha = 2$. [For the series, construct $\mathbf{OA}$, $\mathbf{AB} = \alpha\mathbf{OA}$, $\mathbf{BC} = \frac{1}{2}\alpha\mathbf{AB}$, $\mathbf{CD} = \frac{1}{3}\alpha\mathbf{BC}$, $\mathbf{DE} = \frac{1}{4}\alpha\mathbf{CD}$, etc.]

7. In the plane triangle ABC, whose sides opposite A, B, C, are a, b, c, show by squaring $\mathbf{BC} = \mathbf{AC}-\mathbf{AB}$, that as $a^2 = b^2+c^2-2bc\cos A$; also from
$$V\mathbf{BCCA} = V\mathbf{CAAB} = V\mathbf{ABBC}$$
show that $a:b:c = \sin A : \sin B : \sin C$.

8. From $e^{\theta\epsilon} = \cos\theta+\epsilon\sin\theta$, $e^{\theta\epsilon}\cdot e^{\theta'\epsilon} = e^{(\theta+\theta')\epsilon}$, show that
$$\cos(\theta+\theta') = \cos\theta\cos\theta'-\sin\theta\sin\theta',$$
$$\sin(\theta+\theta') = \sin\theta\cos\theta'+\cos\theta\sin\theta'.$$

9. Show that $(\cos\theta+\epsilon\sin\theta)^n = \cos n\theta+\epsilon\sin n\theta$.

10. Show that $Spq = SpSq+S(Vp\cdot Vq)$, and hence that $\cos\angle pq = \cos\angle p\cdot\cos\angle q+\sin\angle p\cdot\sin\angle q\cdot\cos\angle(Vp\cdot Vq)$.

11. If arc $q = \overset{\frown}{BA}$, arc $p = \overset{\frown}{AC}$, show that the last equation of Ex. 10 is the property
$$\cos a = \cos b\cos c+\sin b\sin c\cos A$$
of the spherical triangle ABC. [Draw $\overset{\frown}{B'A} = \text{arc } Vq$, $\overset{\frown}{AC'} = \text{arc } Vp$.]

12. If α, β, γ, α', β', γ' are vectors from O to the vertices A, B, C, A', B', C' of two spherical triangles on the unit sphere O, where $\alpha' = UV\beta\gamma$, $\beta' = UV\gamma\alpha$, $\gamma' = UV(\alpha\beta)$; then $\alpha = UV\beta'\gamma'$, $\beta = UV\gamma'\alpha'$, $\gamma = UV\alpha'\beta'$, and the two triangles are polar triangles.

13. In Ex. 12 show that $\cos\alpha = -S\beta\gamma$, $\sin\alpha = TV\beta\gamma$, etc.; $\cos A = S(UV\gamma\alpha \cdot UV\alpha\beta) = S\beta'\gamma' = -\cos\alpha'$, etc. Hence $\angle A$, $\angle\alpha'$ are supplements, etc.

14. Show that the equation of Ex. 11 follows from the identity, $-S\beta\gamma = S(\beta\gamma \cdot \gamma\alpha) = S\beta\gamma \cdot S\gamma\alpha + S(V\gamma\alpha \cdot V\alpha\beta)$.

15. From $V(V\gamma\alpha \cdot V\alpha\beta) = -\alpha S\alpha\beta\gamma$, and the similar equations found by advancing the cyclic order α, β, γ, show that we have in the spherical triangle ABC,

$$\sin a : \sin b : \sin c = \sin A : \sin B : \sin C.$$

16. Show that if α, β, γ are coplanar unit vectors, then $\alpha\beta\gamma = -\alpha\beta^{-1} \cdot \gamma =$ (γ turned through the angle from β to α and reversed) $=$ (β rotated 180° about the exterior bisector of the angle between α, γ) $= (\alpha-\gamma)\beta(\alpha-\gamma)^{-1}$.

17. Show that $(V\mathbf{ABCD})^{-1}S\mathbf{AC}V\mathbf{ABCD}$ is the shortest vector from the line AB to the line CD. [Project **AC** upon the common perpendicular to AB, CD.]

18. If α, β, γ be the vector edges about a vertex of an equilateral pyramid (whose edges are unit lengths), then $\beta-\gamma$, $\gamma-\alpha$, $\alpha-\beta$, are the remaining vector edges. Hence show that $S\beta\gamma = S\gamma\alpha = S\alpha\beta = -\frac{1}{2}$, and $V\alpha V\beta\gamma = (-\beta)S\gamma\alpha + \gamma S\alpha\beta = \frac{1}{2}(\beta-\gamma)$. Also show that:

   (a) The face angles are 60°, the area of a face is $\frac{1}{4}\sqrt{3}$, and its altitude is $\frac{1}{2}\sqrt{3}$.

   (b) Opposite edges are perpendicular, and their shortest distance is $\frac{1}{2}\sqrt{2}$.

   (c) The angle between a face and an edge is $\cos^{-1}\frac{1}{3}\sqrt{3}$.

   (d) The angle between two adjacent faces is $\sin^{-1}\frac{2}{3}\sqrt{2}$.

   (e) The volume and altitude of the pyramid are $\frac{1}{12}\sqrt{2}$, $\frac{1}{3}\sqrt{3}$.

19. The cosines of the angles that a vector makes with **i**, **j**, **k**, are called its *direction cosines.* Find the lengths and direction cosines of

$$\mathbf{2i - 3j + 6k}, \mathbf{i + 2j - 2k}, x\mathbf{i} + y\mathbf{j} + z\mathbf{k}.$$

20. Show that the sum of the squares of the direction cosines of a line equals 1.

21. If (l, m, n), (l', m', n') are the direction cosines of two lines, show that $l\mathbf{i} + m\mathbf{j} + n\mathbf{k}$, $l'\mathbf{i} + m'\mathbf{j} + n'\mathbf{k}$, are unit vectors in the directions of the lines, and that if θ be the angle between the lines, then $\cos\theta = ll' + mm' + nn'$; also that $\sin^2\theta = \begin{vmatrix} m & n \\ m' & n' \end{vmatrix}^2 + \begin{vmatrix} n & l \\ n' & l' \end{vmatrix}^2 + \begin{vmatrix} l & m \\ l' & m' \end{vmatrix}^2$, and that the three terms of the second member, respectively divided by $\sin^2\theta$, are the squares of the direction cosines of a line that is perpendicular to the given lines. [Art. 57.]

22. If O be a given origin, then the vector $\mathbf{OP} = \rho = x\mathbf{i} + y\mathbf{j} + z\mathbf{k}$ say, is called the *vector of P with reference to the given origin.* If OX, OY, OZ be axes in the directions of **i**, **j**, **k**, the scalar values of the projections of **OP** upon these axes, *i.e.*, (x, y, z), are called the *coördinates of P with reference to the given axes.* Let the coördinates of the vertices of the pyramid $ABCD$ be, respectively, (8,2,7), (10,6,3), (1,6,3), (9,10,11). Draw this pyramid with reference to a perspective of **i**, **j**, **k**, showing coördinates and vectors. Also:

    (a) Find the vectors and coördinates of the middle points of the edges. [$\mathbf{OM} = \frac{1}{2}(\mathbf{OA} + \mathbf{OB})$, etc.]

    (b) Find the lengths and direction cosines of the edges. [$-\mathbf{AB^2} = AB^2$, etc.]

    (c) Find vectors that bisect the face angles. [$U\mathbf{AC} \pm U\mathbf{AD}$ bisects $\angle CAD$.]

    (d) Find altitudes of the faces and the vectors of their feet. [If L be the foot of the perpendicular from B on AC, then $\mathbf{AL} = \mathbf{AB}^{-1} S\mathbf{ABAC}$, etc.]

    (e) Find the areas of the faces.

    (f) Find the volume and altitudes of the pyramid.

    (g) Find the angles between opposite edges, and their (shortest) distance apart. [Ex. 17.]

    (h) Find the angle between two adjacent faces.

Chapter 4

Equations of First Degree

61. The general equation of first degree in an unknown vector ρ is of the form,

(a) $q_1\rho r_1 + q_2\rho r_2 + \cdots = q,$

where q, q_1, r_1, q_2, r_2, $\cdots$ are known numbers.

This equation may be resolved into two equations by taking the scalar and the vector of each member; and we shall consider these equations separately.

62. Taking the scalar of (a), Art. 61, the term $Sq_1\rho r_1$ becomes, by a cyclic change in the factors, $S \cdot r_1 q_1 \rho$, and this becomes [by dropping the vector $(Sr_1q_1)\rho$, since its scalar is zero] $S(Vr_1q_1 \cdot \rho)$; and similarly for the other terms. Hence if we put $Vr_1q_1 + Vr_2q_2 + \cdots = \delta$, and $Sq = d$, the general scalar equation of first degree in ρ becomes,

(a) $S\delta\rho = d$ or $S\delta(\rho - d\delta^{-1}) = 0.$

One solution of this equation is obviously $\rho = d\delta^{-1}$. This is not the only solution, since by Art. 45, Cor. 3, the second factor may be any vector that is perpendicular to δ. Hence the general solution is $\rho = d\delta^{-1} + V\sigma\delta$, where σ is an arbitrary vector.

63. Hence, draw $\mathbf{OD} = \delta$, take N on the line OD so that $\mathbf{ON} = d\delta^{-1}$, and draw any vector $\mathbf{NP} = V\sigma\delta$ that is perpendicular to the line OD; then $\rho = \mathbf{OP}$ is a solution of the equation $S\delta\rho = d$.

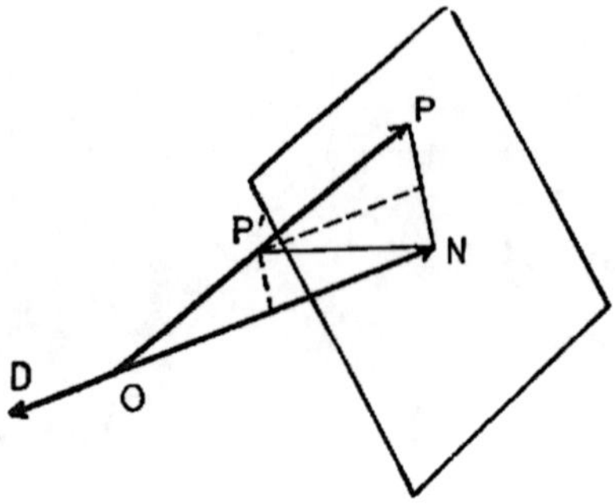

The locus of P is therefore a plane perpendicular to OD at the point N; and this plane is called *the locus of the equation $S\delta\rho = d$, with respect to the origin O.* [The locus is the assemblage of all points that satisfy the equation.]

64. *The vector perpendicular distance from the plane $S\delta\rho - d = 0$ to the point P' (whose vector is ρ') is $\delta^{-1}(S\delta\rho' - d)$, and the corresponding scalar distance measured upon δ is*

$$\frac{-(S\delta\rho' - d)}{T\delta}.$$

For the perpendicular distance of P' is the projection of $\mathbf{NP'} = (\rho' - d\delta^{-1})$, upon $\mathbf{OD} = \delta$.

65. *The locus of the simultaneous equations $S\alpha\rho = a$, $S\beta\rho = b$ is a straight line, viz., the intersection of the two plane loci of these equations taken separately.*

For in order that $\rho = \mathbf{OP}$ may satisfy both equations, P must lie in both planes, and its locus is therefore the intersection, of those planes.

66. The equation $V\delta\rho = \delta'$, or

$$V\delta(\rho - \delta^{-1}\delta') = 0,$$

is a consistent equation only when δ' is perpendicular to δ, since $V\delta\rho$ is always perpendicular to δ. When δ' is perpendicular to δ, then $\delta^{-1}\delta'$

is a vector (Art. 45, Cor. 1), and the general solution of this equation is $\rho = \delta^{-1}\delta' + x\delta$, where x is an arbitrary scalar (Art. 46, Cor.). Hence draw $\mathbf{ON} = \delta^{-1}\delta'$, and $\mathbf{NP} = x\delta$ (any vector parallel to δ), and then $\rho = \mathbf{OP}$ is a solution of the given equation. The locus of P is therefore the straight line through N parallel to δ, and $\mathbf{ON}$ is the perpendicular from the origin upon the line. The equations of Art. 65 take this form by multiplying the first by β, the second by α, and subtracting, remembering that

$$V(V\alpha\beta \cdot \rho) = \alpha S\beta\rho - \beta S\alpha\rho.$$

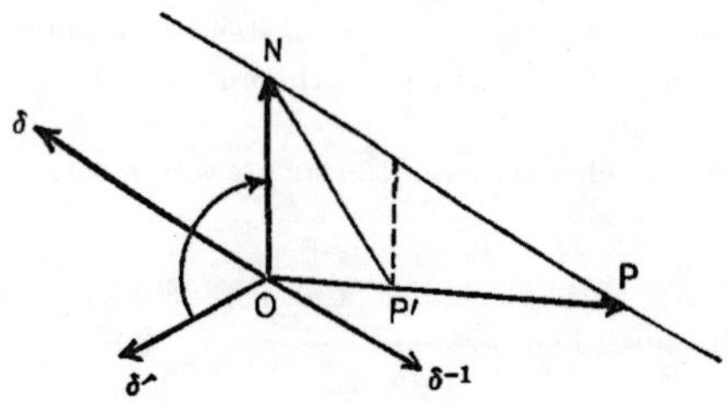

67. *The vector perpendicular distance from the line* $V\delta\rho - \delta' = 0$ *to the point* $\mathbf{P}'$ *is* $\delta^{-1}(V\delta\rho' - \delta')$, *where* $\rho' = \mathbf{OP}'$.

For the required perpendicular distance of P' is the projection of $\mathbf{NP}'$, $= (\rho' - \delta^{-1}\delta')$, perpendicular to δ.

68. *The point of intersection of the three planes* $S\alpha\rho = a$, $S\beta\rho = b$, $S\gamma\rho = c$ *is*

[Art. 56, (j).]
$$\rho = \frac{(aV\beta\gamma + bV\gamma\alpha + cV\alpha\beta)}{S\alpha\beta\gamma}.$$

Examples

1. Find the equation of the locus of a point that moves so that its numerical distances from two fixed points are equal.

2. A point moves so that its scalar distances from two fixed planes are equal; show that its locus is a plane bisector of the diedral angle of the given planes.

3. A point moves so that the sum or difference of its scalar distances from two fixed planes is constant; show that its locus is a plane parallel to the interior or exterior bisector of the diedral angle of the given planes.

4. A point moves so that the ratio of its scalar distances from two fixed planes is constant; show that its locus is a plane.

5. A point moves so that its numerical distances from two intersecting lines are equal; find its locus. [Take the point of intersection as origin.]

6. A point moves so that its numerical distances from three fixed points are equal; find its locus.

   (a) The same with coplanar lines instead of points. [Four straight lines perpendicular to the plane of the lines.]

7. Find the vector of the centre of the sphere whose surface passes through four given points.

8. A point moves so that its tangential distances from two given spheres are numerically equal; find its locus.

9. On the chord $\mathbf{OQ}$ of a given sphere a point P is taken so that $\mathbf{OP} \cdot \mathbf{OQ} = -a^2$; when Q moves round the sphere find the locus of P. [A plane perpendicular to the diameter OD.]

10. The locus of the point P whose coördinates (x, y, z) satisfy $lx+my+nz+d = 0$ is a plane perpendicular to the vector $l\mathbf{i} + m\mathbf{j} + n\mathbf{k}$, at a distance from the origin of $-d/\sqrt{l^2 + m^2 + n^2}$, measured in the direction of this vector. (a) Show that the equation of this plane may be put in the form $x \cos\alpha + y\cos\beta + z\cos\gamma - p = 0$, where p is the perpendicular distance from O to the plane and the cosines are the direction cosines of this perpendicular.

11. Find the perpendicular distance of $P' = (x', y', z')$, from the plane of Ex. 10,

$$[x' \cos\alpha + y' \cos\beta + z' \cos\gamma - p].$$

12. The locus of the point P whose coördinates (x, y, z) satisfy $\frac{(x-a)}{l} = \frac{(y-b)}{m} = \frac{(z-c)}{n}$ is a line parallel to the vector $l\mathbf{i} + m\mathbf{j} + n\mathbf{k}$ through the point (a, b, c). If P satisfy the first two of these three equations, its locus is a plane through the line, perpendicular to the plane of XOY. [If t be the common value of the three ratios, then $\rho = a\mathbf{i} + b\mathbf{j} + c\mathbf{k} + t(l\mathbf{i} + m\mathbf{j} + n\mathbf{k})$.]

13. Find the perpendicular distance of $P' = (x', y', z')$ from the line of Ex. 12.

In the following examples A, B, C, D, P are points whose coördinates are $(8, 2, 7)$, $(10, 6, 3)$, $(1, 6, 3)$, $(9, 10, 11)$, (x, y, z).

14. The equation of the plane through A perpendicular to OD is $S\mathbf{ODAP} = 0$, or $9x + 10y + 11z = 169$.

15. The equation of the plane through AB parallel to CD is $S{\cdot}\mathbf{AP}V\mathbf{ABCD} = 0$, or $2x - 2y - z = 5$.

16. The equation of the plane ABC is $S \cdot \mathbf{AP}V\mathbf{ABAC} = 0$ or $y + z = 9$.

17. Find the perpendicular distance of D from the planes in Exs. 14, 15, 16.

18. The equation of the plane through AB that contains the common perpendicular to AB, CD is

$$S \cdot \mathbf{AP}V(\mathbf{AB}V\mathbf{ABCD}) = 0, \text{ or } 2x + y + 2z = 32.$$

19. The equation of the line through A parallel to OD is $V\mathbf{ODAP} = 0$ or $\mathbf{AP} = t\mathbf{OD}$, or $\frac{(x-8)}{9} = \frac{(y-2)}{10} = \frac{(z-7)}{11}$.

20. The equation of the line AB is $V\mathbf{ABAP} = 0$, or $\mathbf{AP} = t\mathbf{AB}$ or $\frac{(x-8)}{2} = \frac{(y-2)}{4} = \frac{(z-7)}{-4}$.

21. The equation of the common perpendicular to AB, CD is the equation of Ex. 18 and $x + 2y - 2z = 7$.

22. Find the distance of D from the lines in Exs. 19, 20, 21.

23. Find $\mathbf{OD}$ in the form $l\mathbf{OA} + m\mathbf{OB} + n\mathbf{OC}$, and find the ratios in which OD cuts the triangle ABC.

Nonions

69. The vector equation of first degree is

$$(a) \qquad Vq_1\rho r_1 + Vq_2\rho r_2 + \cdots = Vq.$$

To solve this equation we resolve it along $\mathbf{i}, \mathbf{j}, \mathbf{k}$, by multiplying it by these vectors and taking the scalars of the products. We thus find three scalar equations of first degree from which ρ may be immediately found as in Art. 68. Hence (a) has in general one, and only one, solution which corresponds to the intersection of three given planes. [See further Art. 81.]

70. The first member of Art. 69 (a) is a *linear, homogeneous vector function of* ρ; *i.e.*, it is of first degree in ρ, every term is of the same degree in ρ, and it is a vector.

We may denote the operator

$$Vq_1()r_1 + Vq_2()r_2 + \cdots$$

by a single letter, ϕ, so that $\phi\rho, \phi\sigma, \cdots$ denote the vectors that result from putting $\rho, \sigma, \cdots$ in the places occupied by the parenthesis.

71. *The operator* ϕ *is distributive over a sum and commutative with scalars; i.e.*,

$$\phi(x\rho + y\sigma) = x\phi\rho + y\phi\sigma.$$

This is immediately verified by putting $x\rho + y\sigma$ in the places occupied by the parentheses of ϕ and expanding the several terms.

72. We have $\rho = x\alpha + y\beta + z\gamma$, where α, β, γ are given non-coplanar vectors, and x, y, z, are scalars, each of first degree in ρ, as shown in 56(i) with ρ in the place of δ; hence,

$$(a) \qquad \phi\rho = x\phi\alpha + y\phi\beta + z\phi\gamma.$$

The complete operation of ϕ is therefore determined when the three vectors $\phi\alpha, \phi\beta, \phi\gamma$ are known. Since each of these vectors involves three scalar constants (*e.g.*, the multiples of the given non-coplanar vectors α, β, γ, that express it), therefore the value of ϕ depends upon nine scalar constants. The operator ϕ may therefore be called a *nonion*. Scalars and rotators are particular forms of nonions.

NOTE.—It is readily shown that nonions have the same laws of addition and multiplication among themselves as quaternions. Products are not in general commutative. A product

$$(\phi - g_1)(\phi - g_2)(\phi - g_3),$$

where g_1, g_2, g_3 are scalars, is commutative, since ϕ is commutative with scalars by Art. 71. Hence this product multiplies out as if ϕ were a scalar, and is

$$\phi^3 - (g_1 + g_2 + g_3)\phi^2 + (g_2 g_3 + g_3 g_1 + g_1 g_2)\phi - g_1 g_2 g_3.$$

Linear Homogeneous Strain

73. An elastic solid is *subjected to the strain ϕ with respect to an origin O*, when all its particles, A, B, C, etc., are displaced to positions A', B', C', etc., that are determined by $\mathbf{OA'} = \phi\mathbf{OA}$, $\mathbf{OB'} = \phi\mathbf{OB}$, $\mathbf{OC'} = \phi\mathbf{OC}$, etc. In general, any particle P whose vector is $\mathbf{OP} = \rho$ occupies after the strain the position P', whose vector is $\mathbf{OP'} = \phi\rho$. The particle at O is not moved, since its vector after strain is $\phi\mathbf{OO} = \phi 0 = 0$.

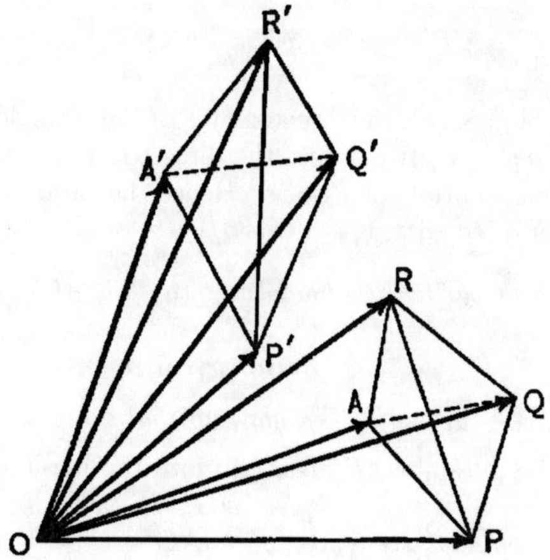

(a) We have, also, $\phi\mathbf{AP} = \mathbf{A'P'}$, etc.

For,

$$\begin{aligned}\mathbf{A'P'} &= \mathbf{OP'} - \mathbf{OA'} = \phi\mathbf{OP} - \phi\mathbf{OA}\\ &= \phi(\mathbf{OP} - \mathbf{OA}) = \phi\mathbf{AP}, \text{ etc.}\end{aligned}$$

74. *A straight line of particles parallel to α is homogeneously stretched and turned by the strain ϕ into a straight line of particles parallel to $\phi\alpha$, and the ratio of extension and turning is $\phi\alpha/\alpha$.*

For let AP be a line parallel to α, and let A, P strain into A', P'. Then, since $\mathbf{AP} = x\alpha$, therefore, by 73*a*, $\mathbf{A'P'} = x\phi\alpha$, and the ratio of extension and turning is $\mathbf{A'P'}/\mathbf{AP} = \phi\alpha/\alpha$.

NOTE.—This property that parallel lengths of the substance strain into parallel lengths and are stretched proportionally, is the physical definition of *linear homogeneous strain.*

75. *A plane of particles parallel to α, β is homogeneously spread and turned by the strain ϕ into a plane of particles parallel to $\phi\alpha$, $\phi\beta$, and the ratio of extension and turning is $V\phi\alpha\phi\beta/V\alpha\beta$.*

For let APQ be a plane parallel to α, β, and let A, P, Q strain into A', P', Q'. Then, since $\mathbf{AP} = x\alpha + y\beta$, $\mathbf{AQ} = x'\alpha + y'\beta$, therefore

$$\mathbf{A'P'} = x\phi\alpha + y\phi\beta, \quad \mathbf{A'Q'} = x'\phi\alpha + y'\phi\beta.$$

By Arts. 59, 55, (e), the directed area of the triangle APQ is $\frac{1}{2}V\cdot\mathbf{AP}\cdot\mathbf{AQ} = \frac{1}{2}(xy' - x'y)V\alpha\beta$, and the directed area of the triangle $A'P'Q'$ is the same multiple of $V\phi\alpha\phi\beta$. Hence the ratio of the extension and turning of directed area is $V\phi\alpha\phi\beta/V\alpha\beta$.

76. *A volume of particles is homogeneously dilated by the strain ϕ in the ratio*

$$S\phi\alpha\phi\beta\phi\gamma/S\alpha\beta\gamma,$$

where α, β, γ are any given non-coplanar vectors.

For let the pyramid $APQR$ strain into the pyramid $A'P'Q'R'$. Then since

$$\mathbf{AP} = x\alpha + y\beta + z\gamma, \quad \mathbf{AQ} = x'\alpha + y'\beta + z'\gamma,$$

$\mathbf{AR} = x''\alpha + y''\beta + z''\gamma$, therefore $\mathbf{A'P'}$, $\mathbf{A'Q'}$, $\mathbf{A'R'}$ have these values with $\phi\alpha$, $\phi\beta$, $\phi\gamma$ instead of α, β, γ. The volume of the pyramid $APQR$ relative to the order AP, AQ, AR of its edges is, by Arts. 60, 56, (c),

$$-\frac{1}{6}S\mathbf{APAQAR} = -\frac{1}{6}\begin{vmatrix} x & y & z \\ x' & y' & z' \\ x'' & y'' & z'' \end{vmatrix} S\alpha\beta\gamma,$$

while that of the strained pyramid is the same multiple of $S\phi\alpha\phi\beta\phi\gamma$. Hence the ratio of dilation of volume is $S\phi\alpha\phi\beta\phi\gamma/S\alpha\beta\gamma$.

77. The ratio of dilation of ϕ is called its *modulus.*

   (a) It is obvious from the signification of the modulus that *the modulus of a product of nonions equals the product of the moduli of the factors; e.g., mod* $\phi\psi = $ *mod* $\phi \cdot$ *mod* ψ.

When mod ϕ is positive, the parts of the volume are in the same order before and after strain. When mod ϕ is negative, the order of the parts is reversed by the strain; *i.e.*, if AP lie on the counter-clockwise side of the plane AQR, then $A'P'$ lies on the clockwise side of $A'Q'R'$, so that the particles along AP have been strained through the particles of the plane AQR. Such a strain is obviously not a physical possibility.

Finite and Null Strains

78. *If an elastic solid which fills all space be subjected to a strain* ϕ, *the strained solid fills all space if mod* ϕ *be finite, and it fills only an indefinite plane or line through the origin or reduces to the origin if mod* ϕ *be zero.*

For if $S\phi\alpha\phi\beta\phi\gamma$ be finite, then $\phi\alpha$, $\phi\beta$, $\phi\gamma$ are non-coplanar vectors, so that

$$\phi\rho(= x\phi\alpha + y\phi\beta + z\phi\gamma)$$

may be made any vector by properly choosing $\rho(= x\alpha + y\beta + z\gamma)$. But if $S\phi\alpha\phi\beta\phi\gamma = 0$, then $\phi\alpha$, $\phi\beta$, $\phi\gamma$ are coplanar vectors or colinear vectors or each zero, so that $\phi\rho$ will be a vector in a given plane or line through O or the vector of O, whatever value be given to ρ.

When mod ϕ is zero, ϕ is called a *null* nonion; and it is called *singly* or *doubly* or *triply* null, according as it strains a solid into a *plane* or a *line* or a *point.* If $\phi\alpha = 0$, then α is called a *null direction* of ϕ.

79. *Null strains, and only null strains, can have null directions; a singly null strain has only one null direction; a doubly null strain has a plane of null directions only; a triply null strain has all directions null.*

For when mod $\phi = 0$, then $\phi\alpha$, $\phi\beta$, $\phi\gamma$ are coplanar or colinear vectors, and we have a relation $l\phi\alpha + m\phi\beta + n\phi\gamma = 0$, *i.e.*, $l\alpha + m\beta + n\gamma$ is a null direction of ϕ. Conversely, if ϕ have a null direction, take one of the three non-coplanar vectors α, β, γ, in that direction, say α, and we have $S\phi\alpha\phi\beta\phi\gamma = 0$, since $\phi\alpha = 0$, and therefore mod$\phi = 0$.

Also, if ϕ have only one null direction, α, then $\phi\beta$, $\phi\gamma$, are not parallel, since $\phi\beta = l\phi\gamma$ makes $\beta - l\gamma$ a second null direction. Since $\rho = x\alpha + y\beta + z\gamma$, therefore $\phi\rho = y\phi\beta + z\phi\gamma$, which is any vector in the plane through O parallel to $\phi\beta$, $\phi\gamma$; hence ϕ is singly null.

But if ϕ have two null directions, α, β, then $\phi\rho = z\phi\gamma$, which is any vector in the line through O parallel to $\phi\gamma$, and therefore ϕ is doubly null. Also, since $\phi(x\alpha + y\beta) = 0$, therefore any direction in the plane of α, β is a null direction of ϕ.

If ϕ have three non-coplanar null directions α, β, γ, then $\phi\rho = 0$ for all values of ρ; *i.e.*, a *triply* null nonion is identically zero.

80. *A singly null nonion strains each line in its null direction into a definite point of its plane; and a doubly null nonion strains each plane that is parallel to its null plane into a definite point of its line.*

For when ϕ is singly null, say $\phi\alpha = 0$, then $x\phi\beta + y\phi\gamma$ is the vector of any point in the plane of ϕ, and all particles that strain into this point have the vectors $\rho = x\alpha + y\beta + z\gamma$, where x is arbitrary, since $\phi\alpha = 0$; *i.e.*, they are particles of a line parallel to α. So, if ϕ is doubly null, say $\phi\alpha = 0$, $\phi\beta = 0$, then any point of the line of ϕ is $z\phi\gamma$, and the particles that strain into this point have the vectors

$$\rho = x\alpha + y\beta + z\gamma,$$

in which x, y are arbitrary; *i.e.*, they are particles of a plane parallel to α, β.

NOTE.—It follows similarly that the strain ϕ alters the dimensions of a line, plane, or volume by as many dimensions as the substance strained contains independent null directions of ϕ, and no more. Hence, a product $\phi\psi$ has the null directions of the first factor, and the null directions of the second factor that lie in the figure into which the first factor strains, and so on; the order of nullity of a product cannot exceed the sum of the orders of its factors, and may be less; etc.

Solution of $\phi\rho = \delta$

81. The solutions of $\phi\rho = \delta$ are, by definition of the strain ϕ, the vectors of the particles that strain into the position whose vector is δ. Hence:

(a) When ϕ is finite, there is one, and only one, solution.

(b) When ϕ is singly null, and δ does not lie in the plane of ϕ, there is no finite solution. Divide the equation by $T\rho$, and make $T\rho$ infinite, and we find $\phi U\rho = 0$; *i.e.*, the vector of the point at infinity in the null direction of ϕ is a solution.

(c) When ϕ is singly null, and δ lies in the plane of ϕ, there are an infinite number of solutions, *viz.*, the vectors of the particles of a line that is parallel to the null direction of ϕ.

(d) When ϕ is doubly null, and δ does not lie in the line of ϕ, there is no finite solution. As in (2) the vectors of the points of the line at infinity in the null plane of ϕ are solutions.

(e) When ϕ is doubly null, and δ lies in the line of ϕ, there are an infinite number of solutions, *viz.*, the vectors of the particles of a plane that is parallel to the null plane of ϕ.

These results correspond to the intersections of three planes, viz.:

(a) The three planes meet in a point.

(b) The three planes parallel to a line.

(c) The three planes meet in a common line.

(d) The three planes parallel.

(e) The three planes coincide.

Derived Moduli of ϕ

82. The ratio in which the nonion $\phi + g$ dilates volume is,

$$\text{mod }(\phi + g) = S(\phi\alpha + g\alpha)(\phi\beta + g\beta)(\phi\gamma + g\gamma)/S\alpha\beta\gamma.$$

This is independent of the values of the non-coplanar vectors α, β, γ in terms of which it is expressed. If g is a scalar, this modulus is an ordinary cubic in g, whose coefficients will therefore depend only upon ϕ. The constant term is mod ϕ, and the coefficients of g, g^2, are called $\mathrm{mod}_1\phi$, $\mathrm{mod}_2\phi$, so that,

$$(a) \qquad \mathrm{mod}\ (\phi + g) = g^3 + g^2\mathrm{mod}_2\phi + g\mathrm{mod}_1\phi + \mathrm{mod}\ \phi.$$

$$[\mathrm{mod}_1\phi = S(\alpha\phi\beta\phi\gamma + \beta\phi\gamma\phi\alpha + \gamma\phi\alpha\phi\beta)/S\alpha\beta\gamma;$$
$$\mathrm{mod}_2\phi = S(\beta\gamma\phi\alpha + \gamma\alpha\phi\beta + \alpha\beta\phi\gamma)/S\alpha\beta\gamma].$$

83. The roots g_1, g_2, g_3, of the cubic

$$\mathrm{mod}\ (\phi - g) = 0$$

are called *the latent roots of* ϕ. We have from 82 (a) with $-g$ in the place of g, and the theory of equations,

$$\mathrm{mod}\ \phi = g_1g_2g_3, \quad \mathrm{mod}_1\phi = g_2g_3 + g_3g_1 + g_1g_2,$$
$$\mathrm{mod}_2\phi = g_1 + g_2 + g_3.$$

(a) *The latent roots of* $\phi - g_1$ *are those of* ϕ *diminished by* g_1.
For the roots of mod $(\phi - g_1 - g) = 0$, are $g = 0$, $g_2 - g_1$, $g_3 - g_1$. *E.g.*, $g = g_2 - g_1$ gives

$$\mathrm{mod}\ [\phi - g_1 - (g_2 - g_1)] = \mathrm{mod}\ (\phi - g_2) = 0.$$

(b) *The order of nullity of* ϕ *cannot exceed the number of its zero latent roots.*
For if ϕ has one null direction α, then $\phi\alpha = 0$ makes mod $\phi = 0$, so that at least one of the latent roots is zero, say g_1; and if ϕ has a second null direction β, then $\phi\alpha = 0$, $\phi\beta = 0$, makes $\mathrm{mod}_1\phi = 0$ or $g_2g_3 = 0$, so that another latent root is zero, etc.

(c) *The order of nullity of* $\phi - g_1$ *cannot exceed the number of latent roots of* ϕ *that equal* g_1. [(a), (b)]

Latent Lines and Planes of ϕ

84. Those lines and planes that remain unaltered in geometrical position by the strain ϕ are called *latent lines and planes of* ϕ.

(a) *The latent directions of* ϕ *are the null directions of* $\phi - g_1$, $\phi - g_2$, $\phi - g_3$, *and* g_1, g_2, g_3 *are the corresponding ratios of extension in those directions.*

For if ρ is any latent direction, and g is the ratio of extension in that direction, then we have $\phi\rho = g\rho$ or $(\phi - g)\rho = 0$. Hence $\phi - g$ is a null nonion, or mod $(\phi - g) = 0$, so that g is a latent root of ϕ; also ρ is a null direction of $\phi - g$.

NOTE.—Since a cubic with real coefficients has at least one real root, therefore a real nonion has at least one latent direction. Also if two roots are imaginary, they are conjugate imaginaries, and the corresponding latent directions must also be conjugate imaginaries.

85. *If* α, β, γ *be the latent directions corresponding to* g_1, g_2, g_3, *then* (β, γ), (γ, α), (α, β) *determine latent planes of* ϕ *in which the ratios of spreading are* $g_2 g_3$, $g_3 g_1$, $g_1 g_2$. *E.g.*,

$$V\phi\beta\phi\gamma = V(g_2\beta \cdot g_3\gamma) = g_2 g_3 V\beta\gamma.$$

Hence, in the general case when the latent roots are all unequal, the latent vectors α, β, γ must form a non-coplanar system, since any two of the latent lines or planes determined by them have unequal ratios of extension, and cannot, therefore, coincide.

(a) *The plane of* $\phi - g_1$ *is the latent plane corresponding to* g_2, g_3.

For $(\phi - g_1)\rho = y(g_2 - g_1)\beta + z(g_3 - g_1)\gamma$, (= plane of β, γ). [The plane and null line of $\phi - g_1$ may be called *corresponding* latents of ϕ.]

The Characteristic Equation of ϕ

86. We have also,

(a) $(\phi - g_1)(\phi - g_2)(\phi - g_3) = 0$. For the first member has the three non-coplanar null directions α, β, γ. [See 80 note, 72 note.]

Conjugate Nonions

87. Two nonions ϕ, ϕ' are conjugate when

(a) $S\rho\phi\sigma = S\sigma\phi'\rho$ for all values of ρ, σ.

When ϕ is known, this determines ϕ' without ambiguity. Thus, put $\sigma = \mathbf{i}, \mathbf{j}, \mathbf{k}$, in turn, and we have by Art. 57 (b),

$$\phi'\rho = -\mathbf{i}S\rho\phi\mathbf{i} - \mathbf{j}S\rho\phi\mathbf{j} - \mathbf{k}S\rho\phi\mathbf{k}.$$

Conversely, this function satisfies (a), for we have $S\sigma\phi'\rho = S\rho\phi(-\mathbf{i}S\mathbf{i}\sigma - \mathbf{j}S\mathbf{j}\sigma - \mathbf{k}S\mathbf{k}\sigma) = S\rho\phi\sigma$.

88. From this definition of conjugate strains we have

(a) $(a\phi + b\psi)' = a\phi' + b\psi'$; $(\phi\psi)' = \psi'\phi'$.

(b) $(Vq()q)' = Vq()p$, $[\alpha S\beta()]' = \beta S\alpha()$.

$$\begin{aligned} E.g., S\sigma(\phi\psi)'\rho &= S\rho\phi\psi\sigma = S\rho\phi(\psi\sigma) \\ &= S\psi\sigma\phi'\rho = S\sigma\psi'\phi'\rho, \end{aligned}$$

and therefore $(\phi\psi)' = \psi'\phi'$. [If $S\sigma(\alpha - \beta) = 0$ for all values of σ, then $\alpha - \beta = 0$, since no vector is perpendicular to every vector σ. Hence, comparing the first and last member of the above equation, we have $(\phi\psi)'\rho = \psi'\phi'\rho$.]

89. *Two conjugate strains have the same latent roots and moduli, and a latent plane of one is perpendicular to the corresponding latent line of the other.*

For since $(\phi - g_1)\alpha = 0$, therefore

$$0 = S\rho(\phi - g_1)\alpha = S\alpha(\phi' - g_1)\rho,$$

and therefore $\phi' - g_1$ is a null nonion whose plane is perpendicular to α. Hence g_1 is a latent root of ϕ', and the latent plane of ϕ' corresponding to $\phi' - g_1$ is perpendicular to the latent line of ϕ corresponding to $\phi - g_1$. [Art. 85, (a).]

Self-conjugate Nonions

90. A nonion ϕ is *self-conjugate* when $\phi' = \phi$ or when $S\rho\phi\sigma = S\sigma\phi\rho$ for all values of ρ, σ. In consequence of this relation a self-conjugate strain has only six scalar constants, three of the nine being equal to three others, *viz.*,

$$S\mathbf{i}\phi\mathbf{j} = S\mathbf{j}\phi\mathbf{i}, \quad S\mathbf{i}\phi\mathbf{k} = S\mathbf{k}\phi\mathbf{i}, \quad S\mathbf{j}\phi\mathbf{k} = S\mathbf{k}\phi\mathbf{j}.$$

91. A self-conjugate strain has by Art. 88 three mutually perpendicular latent directions, and conversely, if ϕ have three mutually perpendicular latent directions, $\mathbf{i}$, $\mathbf{j}$, $\mathbf{k}$, corresponding to latent roots a, b, c, then

$$\phi\rho = -a\mathbf{i}S\mathbf{i}\rho - b\mathbf{j}S\mathbf{j}\rho - c\mathbf{k}S\mathbf{k}\rho,$$

which is self-conjugate. [68 b.]

92. *A real self-conjugate strain has real latent roots.*

For let $\alpha' = \alpha + \beta\sqrt{-1}$, $\beta' = \alpha - \beta\sqrt{-1}$ be latent directions corresponding to conjugate imaginary roots a, b of a real nonion ϕ; then, if ϕ is self-conjugate, we have

$$S\alpha'\phi\beta' = S\beta'\phi\alpha' = bS\alpha'\beta' = aS\alpha'\beta',$$

or, since a, b are unequal, therefore $S\alpha'\beta' = 0$; but this is impossible, since $S\alpha'\beta' = \alpha^2 + \beta^2$, a negative quantity. Therefore ϕ is not self-conjugate if it has imaginary latent roots.

93. A nonion ϕ is *negatively self-conjugate* when $\phi' = -\phi$, or when $S\sigma\phi\rho = -S\rho\phi\sigma$. Such a nonion has therefore only three scalar constants, since $S\mathbf{i}\phi\mathbf{i} = -S\mathbf{i}\phi\mathbf{i}$ shows that $S\mathbf{i}\phi\mathbf{i} = 0$, and similarly, $S\mathbf{j}\phi\mathbf{j} = 0$, $S\mathbf{k}\phi\mathbf{k} = 0$, while the other six constants occur in negative pairs

$$S\mathbf{i}\phi\mathbf{j} = -S\mathbf{j}\phi\mathbf{i}, \text{ etc.}$$

(a) The identity $S\rho\phi\rho = 0$ gives (by putting $\rho = x\mathbf{i} + y\mathbf{j} + z\mathbf{k}$ where x, y, z are arbitrary) all the above relations between the constants of ϕ, and is therefore the sufficient condition that ϕ is negatively self-conjugate. It shows that $\phi\rho$ is perpendicular to ρ or that $\phi\rho = V\epsilon\rho$, where ϵ must be independent of ρ since $\phi\rho$ is linear in ρ.

94. *Any nonion ϕ may be resolved into a sum of a conjugate and a negatively self-conjugate nonion in only one way.*

For if $\phi = \overline{\phi} + \psi$, where $\overline{\phi'} = \overline{\phi}$, $\psi' = -\psi$, then $\phi' = \overline{\phi} - \psi$, and adding and subtracting, we have $\overline{\phi} = \frac{1}{2}(\phi + \phi')\psi = \frac{1}{2}(\phi - \phi')$, and

$$\phi\rho = \frac{1}{2}(\phi + \phi')\rho + \frac{1}{2}(\phi - \phi')\rho = \overline{\phi}\rho + V\epsilon\rho. \tag{a}$$

To find ϵ in terms of the constants of ϕ, we have $\rho = -\mathbf{i}S\mathbf{i}\rho - \mathbf{j}S\mathbf{j}\rho - \mathbf{k}S\mathbf{k}\rho$, and therefore

$$\phi\rho = -\phi\mathbf{i}S\mathbf{i}\rho - \text{ etc.}$$

[88 *b*.]

$$\phi'\rho = -\mathbf{i}S\rho\phi\mathbf{i} - \text{ etc.}$$

Hence

$$\begin{aligned}\frac{1}{2}(\phi - \phi')\rho &= \frac{1}{2}(\mathbf{i}S\rho\phi\mathbf{i} - \phi\mathbf{i}S\mathbf{i}\rho) + \text{ etc.}\\ &= \frac{1}{2}V\cdot(V\mathbf{i}\phi\mathbf{i})\rho + \textit{ etc.} = V\epsilon\rho,\end{aligned}$$

and therefore

$$\epsilon = \frac{1}{2}V(\mathbf{i}\phi\mathbf{i} + \mathbf{j}\phi\mathbf{j} + \mathbf{k}\phi\mathbf{k}). \tag{a}$$

Examples

1. Find the equation of a sphere whose centre is $A(\mathbf{OA} = \alpha)$ and radius α.
2. Show that the square of the vector tangent from the sphere of Ex. 1 to P' is $(\rho' - \alpha)^2 + a^2$.

3. Find the locus of the point P such that PP' is cut in opposite ratios by the sphere of Ex. 1; show that it is the plane of contact of the tangent cone from P' to the sphere and is perpendicular to AP'.

4. Let P' be any point on the sphere A of Ex. 1, and take P on OP' so that $\mathbf{OP}\cdot\mathbf{OP}'+c^2=0$; find the locus of P. [P, P' are called *inverse* points with respect to O, and the locus of P is the *inverse* of the given sphere A. It is a sphere with centre A' on OA, or a plane perpendicular to OA if the given sphere A pass through O.]

5. Show that the inverse of a plane is a sphere through O.

6. Show that the general scalar equation of second degree is $S\rho\phi\rho+2S\delta\rho+d=0$, where ϕ is a self-conjugate nonion.

7. Show that $S\rho\phi\rho=0$ is the equation of a cone with vertex at O.

8. Show that the line $\rho=\alpha+x\beta$ cuts the quadric surface of Ex. 6 in two points; apply the theory of equations to determine the condition that this line is a tangent to the surface, or an element of the surface, or that it meets the surface in one finite point and one point at infinity, or that the point whose vector is α lies midway between the points of intersection.

9. Show that the solution of $\phi\rho+\delta=0$ is the vector of a centre of symmetry of the quadric surface of Ex. 6. Hence classify quadric surfaces as *central*, *non-central*, *axial*, *non-axial*, *centro-planar*.

10. Show that the locus of the middle points of chords parallel to β is a diametric plane perpendicular to $\phi\beta$.

11. Show that an axial quadric is a cylinder with elements parallel to the null direction of its nonion ϕ.

12. Show that a non-axial quadric is a cylinder with elements parallel to the null plane of its nonion ϕ and perpendicular to its vector δ.

13. Show that a centro-planar quadric consists of two planes parallel to the null plane of its nonion ϕ.

14. Show that the equation of a central quadric referred to its centre as origin is $S\rho\phi\rho+1=0$. Show that the latent lines and planes of ϕ are axes and planes of symmetry of the quadric; also that $\phi\rho$ is perpendicular to the tangent plane at the point whose vector is ρ. (a) Show that the axes and planes of symmetry of the general quadric are parallel to the latent lines and planes of ϕ.

15. Show that if $\psi^2 = \phi$, then the equation of the central quadric is $(\psi\rho)^2 + 1 = 0$; and that therefore the quadric surface when strained by ψ becomes a spherical surface of unit radius.

16. Show that if g, α are corresponding latent root and direction of ϕ, then g^n, α are the same for ϕ^n. Find the latent lines and planes, the latent roots and moduli of the following nonions and their powers:

    (a) $(a\alpha S\beta\gamma\rho + b\beta S\gamma\alpha\rho + c\gamma S\alpha\beta\rho)/S\alpha\beta\gamma$.

    (b) $[a\alpha S\beta\gamma\rho + (a\beta + b\alpha)S\gamma\alpha\rho + (c\gamma S\alpha\beta\rho]/S\alpha\beta\gamma$.

    (c) $[a\alpha S\beta\gamma\rho + (a\beta + b\alpha)S\gamma\alpha\rho + (a\gamma + c\beta)S\alpha\beta\rho]/S\alpha\beta\gamma$.

    (d) $V\epsilon\rho$, $q\rho q^{-1}$.

17. Show that the latent roots of $e\rho - fV\alpha\rho\beta$ $(f > 0, T\alpha = T\beta = 1)$ are $e + f$, $e + fS\alpha\beta$, $e - f$, corresponding to latent directions $\alpha + \beta$, $V\alpha\beta$, $\alpha - \beta$; and that this is therefore a general form for self-conjugate nonions. Determine the latent directions and roots in the limiting case when $\alpha = \beta$, or $-\beta$ or $f = 0$.

18. Show that the nonion of Ex. 16 takes the form $b\rho - f(\alpha S\beta\rho + \beta S\alpha\rho)$, where b is the mean latent root.

19. Substitute the nonion of Ex. 18 for ϕ in Ex. 6 and show that the quadric surface is cut in circles by planes perpendicular to α or β. When is the surface one of revolution?

20. If the conjugate of a nonion is its reciprocal, and the modulus is positive, then the nonion is a rotation; and conversely every rotation satisfies this condition. [If R, R^{-1} are conjugate nonions, then $\rho^2 = S\rho R^{-1}R\rho = SR\rho R\rho = (R\rho)^2$; *i.e.*, $TR\rho/\rho = 1$. Also $S\rho\sigma = S\rho R^{-1}R\sigma = SR\rho R\sigma$ and therefore the angle between $\rho, \sigma = \angle$ between $R\rho, R\sigma$. Therefore R strains a sphere with centre O into another sphere with centre O in which the angles between corresponding radii are equal and their order in space is the same, since mod R is positive. Hence the strain is a rotation.]

    (a) Show that $(R\phi R^{-1})^n = R\phi^n R^{-1}$.

21. Show that $\phi'\phi$ is self-conjugate, and that its latent roots are positive, and that therefore there are four real values of ψ that satisfy $\psi^2 = \phi'\phi$, mod ψ = mod ϕ. [Let $\phi'\phi\mathbf{i} = a\mathbf{i}$; then $a = -S\mathbf{i}\phi'\phi\mathbf{i} = (T\phi\mathbf{i})^2$.]

22. If $\phi = R\psi$, where ψ is the self-conjugate strain $\sqrt{\phi'\phi}$, then R is a rotation. So $\phi = \chi R$, where $\chi = R\psi R^{-1} = \sqrt{\phi\phi'}$.

23. Show that $\phi' \cdot V\phi\beta\phi\gamma = V\beta\gamma \cdot \text{mod}\ \phi$. [56 *j.*]

24. Show that the strain $\phi\rho = \rho - a\alpha S\beta\rho$, where α, β, are perpendicular unit vectors, consists of a shearing of all planes perpendicular to β, the amount and direction of sliding of each plane being $a\alpha$ per unit distance of the plane from O.

25. Determine ψ and R of Ex. 22 for the strain of Ex. 24, and find the latent directions and roots of ψ.